A Natural History of Shells

A Natural History of

SHELLS

BY

Geerat J. Vermeij

With a new preface by the author

PRINCETON UNIVERSITY PRESS

PRINCETON AND OXFORD

First published by Princeton University Press in 1993
New Princeton Science Library paperback,
with a new preface by the author, 2021
Paperback ISBN 978-0-691-22924-9

Library of Congress Control Number: 2021934445

This book has been composed in Adobe Baskerville
with Avant Garde display
Text designed by Jan Lilly

Cover image: iStock

CONTENTS

PART III: THE DIMENSION OF TIME

Ever since I was a young boy in the Netherlands, I have liked shells. Like the lush plants with their fragrant flowers growing in the wet meadows of the polders surrounding our town of Gouda, and the songs of chiffchaffs and blackbirds in the pine woods at my boarding school for the blind in Huizen, the shells I gathered with my parents and brother Arie at the wide beach in Scheveningen stimulated in me an intense feeling of beauty. These simple productions of nature opened up an almost magical world of pleasure and surprise, and instilled an abiding curiosity about living things, a welcome distraction from the unhappiness of being away from home for long periods in the noisy and seemingly uncaring human environment of the boarding school.

As I grew older and came to know the almost unfathomable elegance of tropical shells, my love blossomed into an enthusiasm for all things scientific. With the help of everyone in my close-knit family, I read everything I could about the sea, about fossils, plants, molluscs, and outer space. I began to ask questions about the many puzzles I confronted with my eager fingers and ears. Why were those tropical shells so ornate compared with the cold-water shells of my earlier days? Why were there trilobites in ancient seas but not in today's? Arie and I grew beans, radishes, and sunflowers from seeds, and everyone at home read to me, drew tactile maps and illustrations, and described the many things I could not see. I began to collect shells in earnest, and also seeds, pressed plants, feathers, and even wood samples. We dreamed of faraway places with captivating names, and I was immediately drawn to the Latin and Greek cadences of the scientific names of the molluscs and other creatures that we encountered in books.

The scientific frame of mind, soon to be informed by the notion that living things evolve through natural and knowable processes, suited me perfectly. I was intrigued by the big ideas of science, as well as by the elegance of mathematics and the periodic table. Shells became my entry point into the realm of ideas, of understanding how nature works and changes.

The study of shells was at first a worthy end in itself, but it gradually enabled me to realize that it was much more. It opened the door to think not just about the history of life on earth, but even about the human realm, about the economic principles that govern evolution and the ecosystems we inhabit and create. The molluscan makers of shells, like the rest of us, must make a living in a world of opportunities and limitations, of enabling factors and competition, of selective processes and cooperative arrangements, all in the face of unpredictable calamity. Through their adaptive shapes and the individual life events recorded in their shells, molluscs chronicle the conditions of life now and in the distant past. The beauty of nature, the satisfaction of uncovering the deep regularities of history, and the thrill of perceiving connections among seemingly unrelated phenomena contribute for me to a profound, emergent meaning of life, one centered on a love of science and of like-minded scholars and thoughtful friends, not least my wonderful wife, Edith.

My goal in this book is to communicate this enthusiasm and some of the insights I have gained from molluscan shells, so that others may also appreciate and continue to build on this scientific enterprise. Knowledge and understanding are like money; they are of little use if they are hoarded or kept to oneself. Only when they are shared do knowledge and the quest for it become truly meaningful. My hope is that this book will open doors to the scientific way of knowing and the wonders of the natural world, and that it will enrich others' lives just as the study of shells has enriched mine.

A good deal more is known about shells now than when this book was first published. Many of the scientific names used in this book have changed, mainly at the genus level but occasionally also at the level of species. New work, including my own, has substantially increased our understanding of many of the topics and questions dealt with in this book. For example, we now know that the cap-shaped or slipper-like limpet shell form in gastropods arose at least fifty-four times independently. Innovations such as the labral tooth, a protrusion on the edge of the lip of the shell that speeds up predation by gastropods on hard-shelled prey, occur repeatedly in some evolutionary branches and never in others. It has also become increasingly clear that shell form is influenced not only by the selective effects of predation and other agents of natural selection but also by the forces to which molluscs are subjected and that they themselves create through their activities. Thanks to great advances in the study of DNA sequences and the fossil record, the evolutionary tree of molluscs is now far better known, Nevertheless, most of what is in this book remains unchanged; it has only been more thoroughly documented and placed on a more secure evolutionary footing.

Although all of us scientists make individual contributions, and many of us often work alone or with one or two collaborators, our discoveries could never have come about without the help and encouragement of thousands of people and institutions. I am profoundly grateful to my family, to the schools and teachers in the Netherlands and the United States, and for the support and stimulation I received while studying at Princeton University and Yale University. Without dedicated field assistants, it would have been impossible for me to conduct fieldwork, often in very challenging places like tropical coral reefs, rainforests, muddy mangrove swamps, wave-swept rocky shores, and the frigid Aleutian Islands. I have benefited greatly from research grants awarded by the National Science Foundation and the National Geographic Society, and from the intellectual freedom implicit in awards from the John Simon Guggenheim Foundation and the John D. and Catherine T. MacArthur Foundation. I have had the luxury of teaching and pursuing research of my choosing thanks to the academic liberty that is the norm at the University of Maryland at College Park and the University of California, Davis. Curators at natural history museums the world over enabled me to study the collections in their care and to dig up obscure works in their libraries. Bettina Dudley; Janice Cooper; Alyssa Henry; Tracy Thomson; my wife, Edith Zipser; and others have read tens of thousands of scientific works to me as I transcribed them into my Braille library. Thoughtful reviewers and editors have helped me to improve my publications, and they give me confidence that the peer-review system, a fundamental component of the trust that science enjoys and requires, is healthy and as important as ever. The act of teaching, together with the many stimulating conversations with students and colleagues, has been essential to my growth as a scholar. Finally, but not least, I am grateful to the composers, performers, and broadcasters of classical music, which brings me great joy and contentment and which accompanies me daily as I write.

No living thing exists and evolves in isolation. Our human lives, scientific and otherwise, are likewise shaped and enriched by our interactions with the world and with the people whom we love, trust, influence, and depend on. It is a privilege to wonder, to discover, and to communicate, to ask questions, to consider evidence, and to arrive at a coherent worldview of how things work. Above all, there is beauty to be cherished, understood, and perpetuated.

A Natural History of Shells

CHAPTER 1

Shells and the Questions of Biology

Few works of architecture can match the elegance and variety of the shells of molluscs. Beauty is reason enough to appreciate and study shells for their own sake, but shells offer much more. As molluscs grow, they enlarge their shells little by little, and in doing so inscribe in their shells a detailed record of the everyday events and unusual circumstances that mark their lives. Moreover, the fossil record that chronicles the history of life is replete with the shells of extinct species. We can therefore learn about the conditions of life and death of molluscs not just in our own world, but in the distant past. The sizes, shapes, and textures of shells inform us about the way skeletons are built and how animals respond to the hazards around them.

The molecular biologist Sidney Brenner once observed that there are three fundamental questions we can ask about a biological structure.* How does it work? How is it built? How did it evolve? These questions apply to structures at all levels of the organic hierarchy, from proteins to cells to whole animals, populations, and ecosystems.

The first of Brenner's questions is one of mechanics and effectiveness of design. What is the relationship between structure and function, and how well does the structure work under given conditions? What are the mechanical principles and the circumstances that dictate the possibilities and limitations of adaptive design?

The second question deals with the rules of biological design. What are the rules by which individual organisms develop and grow, and how do they work? What limits do they impose on the diversity of forms encountered in nature? How, and under what circumstances, can change be brought about within the established pattern as defined by the rules? What happens when the rules are broken or relaxed, and when can this occur?

The third line of inquiry is historical. All living and fossil species trace their ancestry back to a single entity in the incredibly distant past. What was the course of this evolution, and what factors were important in charting it? To what extent does a species bear the stamp of history, and how much do its characteristics reflect the conditions

*Horace F. Judson, *The Eighth Day of Creation*, Simon and Schuster, New York; p. 218.

in which it finds itself? When and how does evolutionary change occur, and how is this change constrained by the rules of construction and by the environment in which organisms live?

In the context of shells, these three fundamental questions can be effectively framed in economic terms. We can think of shells as houses. Construction, repair, and maintenance by the builder require energy and time, the same currencies used for such other life functions as feeding, locomotion, and reproduction. The energy and time invested in shells depend on the supply of raw materials, the labor costs of transforming these resources into a serviceable structure, and the functional demands placed on the shell. For secondary shell-dwellers, which generally cannot enlarge or repair their domiciles, the quantity and quality of housing depend on the rate at which shells enter the housing market and on the ways and rates of shell deterioration. The words "economics" and "ecology" are especially apt in this context, for both are derived from the Greek *oikos*, meaning house. In short, the questions of biology can be phrased in terms of supply and demand, benefits and costs, and innovation and regulation, all set against a backdrop of environment and history.

Shells are, of course, more than houses. For many molluscs and most secondary occupants, they are also vehicles, which are often specifically adapted to various modes of locomotion such as crawling, leaping, swimming, and burrowing. Moreover, shells in some instances function as traps for prey and would-be intruders, as offensive weapons of attack, as signals for attracting mates, and even as greenhouses for culturing plant cells that help feed the animal. The various functional demands are apt to be incompatible with each other. The architecture of any one shell reflects not only the compromises among these functional requirements, but also the costs of construction and maintenance, the rules governing growth, and the mark of evolutionary ancestry. Just as the houses of people vary greatly from place to place and over the course of history, so the shells of molluscs bear the marks of geography and time. Costs of construction vary according to geography and habitat; so do the kinds and abundances of predators, the availability of food, the rate of growth, and any number of other factors important in the lives of shell-bearers. Ecologists who wish to understand how population sizes of living species are regulated may be content to document these variations in the biosphere today, but for the evolutionary biologist interested in chronicling the economic history of life, it becomes essential to determine how costs, benefits, and resources have varied over the course of geologic history, and to infer how the course of evolution has been influenced by the interplay between the everyday economic forces and the much less frequent large-scale changes in climate and tectonics that have affected the planet. Such evolutionary insights will be important in attempts to forecast and manage biological change as humans extend their control over the biosphere.

An economic treatment of biology is, of course, not new. Cost-benefit analysis has pervaded much of the literature in evolutionary ecology for the last 25 years. My approach, however, differs from that of most others who have concerned themselves with the economy of nature. The prevailing doctrine has been that organisms are optimally designed to maximize the intake of resources while minimizing costs and risks. If organisms fall short of the optimum, an appeal is generally made to factors that are either unknown or unmeasured. The underlying assumption is always that natural selection—the process by which genes conferring higher survival or reproduction are favored—produces the best design possible given the circumstances in which a population lives.

I find this point of view profoundly antievolutionary. When individual organisms vie for resources—mates, food, living quarters, safe places, and the like—the winner is superior in some way to the loser, as ultimately measured in survival and reproduction. Sometimes being better means being very good indeed, but in other circumstances success is achievable with what, in absolute terms, appears to be only a modest effort. By thinking of selection as favoring a better organism rather than as favoring the best organism, we are at once dismissing the notion of an adaptational ideal. Optimality implies a directedness, even a purpose, for whose existence there is no evidence whatever. Humans can think up strategies and tactics in order to improve their lives or to enhance their own causes, but natural selection acts only in the here and now and is therefore fundamentally different from long-range purposeful planning. Evolutionary change can track environmental change but cannot forecast or plan for it.

The order of topics in this book departs slightly from Brenner's sequence of questions, because it recapitulates the pathway by which I came to the study of shells. From my earliest acquaintance with shells in the Netherlands, I was drawn to the regularity of form that even the simplest and most ordinary shells displayed. Having picked up only empty shells, I saw them as abstract objects. The fact that animals built them and inhabited them was unknown to me. The first part of the book is therefore an exercise in geometry. From a description of shell form, I shall proceed to the rules of construction and arrive at a model that explains some of the basic features of shell architecture.

In the Netherlands I had become accustomed to the chalky and rather sloppily ornamented clam shells that washed up in great profusion on the North Sea beaches. Shortly after coming to the United States, I had the great fortune to be in Mrs. Colberg's fourth-grade class in Dover, New Jersey. The windowsills of her classroom held a display of the shells she had gathered on her travels to the west coast of Florida. My first glimpse of these shells is deeply etched in memory. Here were elegantly shaped clam and snail shells, many adorned with neatly arranged ribs, knobs, and even spines. Not only were the shell interiors impossibly smooth to the touch, but the olive and cowrie shells were externally so polished that I was certain someone had varnished them. The contrast with the drab chalky shells from the Netherlands was remarkable. Why, I wondered, were warm-water shells so much prettier than the northern shells? When a classmate brought in some shells from the Philippines, which were even more spectacular in their fine sculpture and odd shapes, my curiosity was aroused even more. I resolved to begin collecting shells and to read as much as I could find about them.

The geography of shell form has remained a matter of interest for me ever since. It forms the point of departure for the rest of the book. I begin by examining the economic costs of shell construction, and proceed by asking how these costs vary with geography and habitat. Next I review what we know about how shells work, and ask how the factors with which shell-bearers must cope vary with latitude and other geographic and habitat gradients. Differences in shell architecture among molluscan assemblages from different oceans lead into an exploration of how historical factors have conspired to make molluscs and other animals in some parts of the world functionally more specialized than in others. This inquiry, in turn, expands into an architectural and functional history of molluscs from the time of the first appearance of the group in the Early Cambrian, some 530 million years ago, to the present. I close with some suggestions about what we can learn about our own species from the lessons of the history of life.

PART I

The Rules of Shell Construction

Themes and Variations: The Geometry of Shells

The shells of molluscs derive much of their aesthetic appeal from the regularity of their form. From the platelike valves of scallops to the tightly wound needle-shaped shells of auger snails, shells are endless variations on a geometric theme in which an expanding figure sweeps out a curved or spirally coiled hollow edifice. Because shells are growing structures built by animals, an appreciation of how these variations are brought about must rest on an understanding of how shells grow. Once we know these rules of growth and form, we can ask why certain shapes that are compatible with the rules are rarely or never encountered in nature.

In this chapter I first introduce the animals that build the shells and then explore the principles of growth and form that govern shell construction. I arrive at an interpretation of shell geometry that departs from a tradition dating back almost 200 years. In chapter 3, I examine shell construction from the perspective of an economist. Together, the principles of geometry and economics provide the basis for an inquiry into how shells work, a subject treated in Part II.

An Introduction to Molluscs

Molluscs make up the second largest major group (or phylum) in the animal kingdom. There are more than 50,000 living species, and many thousands of fossil species that have become extinct since the group first appeared about 550 million years ago. Today, molluscs live nearly everywhere on land, in fresh water, and in the sea, from the polar zones to the tropics, and from the highest mountain peaks to the deepest ocean trenches. Every imaginable way of life is practiced by one or another representative of the phylum. Many adult clams and snails live permanently attached to other objects and filter food particles out of the surrounding water. At the opposite extreme, squids, whose swimming speeds match those of fish, are predators that use vision to locate their prey. There are herbivores and carnivores, parasites and mud-eaters, giant clams that garden green plants in their tissues, deep-sea molluscs that culture bacteria, and ship-worms that bore into wood. The venom of fish-eating species of *Conus* is so potent that shell collectors who have pocketed the coveted shells of these snails have occasionally been fatally stung.

In order to make sense out of the bewildering diversity of life, biol-ogists have devised an elaborate hierarchical scheme of classification. Lowest in the hierarchy is the species. This is what we think of as a kind of animal or plant. Its members are individuals that are geneti-cally compatible with one another in nature. Each species is given a two-part scientific name. The first name is that of the genus, the next level in the hierarchy, to which the species belongs; the second is the so-called specific name. These are supposed to be Latin or Greek words, but this rule is often rather loosely interpreted by those who name and describe species. The snail *Conus coronatus* has an apt name meaning "crowned cone," referring to the fact that the cone-shaped shell has a crownlike circle of knobs at its wide end; but a name like *Schwartziella newcombei* (literally Newcombe's little Schwartz) does not evoke the characters of the species and hardly qualifies as either Latin or Greek.

A genus contains one or more species that share a large number of characteristics and differ from other genera. Moving up in the hierar-chy, genera are grouped into families, which again are united by a set of characteristics that its members have in common. Each family name is based on the name of one of the component genera, and ends in the suffix "-idae."

Families are grouped into orders, orders into classes, and classes into phyla. At the top of the hierarchy stands the kingdom.

To take one example, the tiny European snail *Caecum trachea* belongs to the family Caecidae, the order Mesogastropoda, the class Gastropoda, the phylum Mollusca, and the kingdom Animalia. There are also ranks between these basic units: subgenera, subfamilies, superfamilies, suborders, subclasses, subphyla, and so on. *Caecum*, for example, belongs to the superfamily Truncatelloidea and the subclass Prosobranchia. The task of those who classify and categorize molluscs can perhaps best be summed up in the slogan, "You seek 'em, we'll file 'em."

Ideally, each unit of classification corresponds to an evolutionary branch (or clade). A clade has a single common ancestor and con-tains all the descendants of that common ancestor. The unit is distin-guished by characteristics not shared with other clades, and can be evolutionarily related to other clades by characters it has in common with them by virtue of their common ancestry. In practice, many categories in the classificational scheme do not meet these criteria. As a result, the system of molluscan classification is undergoing revision and refinement constantly.

The phylum Mollusca evidently comprises two great branches. Most members of both subphyla possess shells, but it is likely that shells evolved independently in the two groups. Shells are external skeletons built of the mineral calcium carbonate by the edge of the mantle, a skirtlike cover that surrounds the internal organs of the animal. The mantle edge deposits thin layers of mineral along the growing edge of the shell and therefore enlarges the shell at its open end.

The smaller of the two subphyla contains two classes, the Polyplacophora (chitons) and the wormlike Aplacophora. The shell of chitons is an eight-part shield covering the soft parts from above (fig. 2.1). The eight valves are arrayed in a line from the front (anterior) to the back (posterior) of the animal. The front edges of one valve are extended beneath the valve in front, and they are surrounded by a flexible girdle. The chiton crawls on its foot, a muscular organ that clings to hard surfaces by means of mucus. Most chitons eat seaweeds, but others nourish themselves with sponges and other sedentary animals, and North Pacific members of the genus *Placiphorella* trap small crustaceans beneath the front portion of the shell, which can be lifted up and then rapidly brought down over the prey. The Aplacophora are wormlike animals evidently derived from shell-bearing chitons, but because they lack external hard parts, they will not be considered further in this book.

The second subphylum (Conchifera) is usually divided into six classes. The stem "class" from which all members of this great group are believed to be derived is a grab bag of primitive molluscs collectively referred to as the Monoplacophora. Their shell consists of a single piece, or univalve, which varies from cap-shaped (limpetlike) to coiled. Until the 1950s, it was thought that all Monoplacophora were extinct, having disappeared about 400 million years ago during the Devonian period. In one of the more dramatic zoological discoveries of the twentieth century, however, deep-sea monoplacophorans were found off the coast of Peru, and have since been recognized from deep waters in most of the world's oceans. All are limpetlike members of the family Neopilinidae.

The largest molluscan class is the Gastropoda, the snails. Again the shell is a univalve (fig. 2.2), whose opening (aperture) is often covered by a door (operculum) when the soft parts are retracted into the

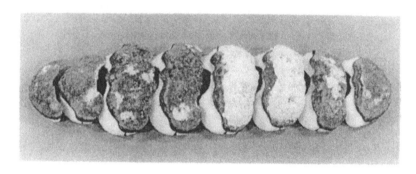

Fig. 2.1. Chitons. Top. A whole *Tonicella lineata* from Chuginadak Island. Aleutian Islands. Alaska, with all eight valves connected together as a shield over the soft body. The specimen is 38 mm long. Bottom. Separated valves of *Acanthopleura echinata* from Montemar, Chile. The widest valve is 48 mm wide.

shell. Most snails are marine, but snails have explored every mode of life that is to be found among the Mollusca. They are the only land molluscs, and together with clams have diversified extensively in fresh water. In size they range from less than 1 mm in adult diameter to a length of 1 m.

By far the largest molluscs, however, belong to the class Cephalopoda. Giant squids of the genus *Architeuthis* reach a length of 10 m or more, and even the smallest cephalopods (2 cm long) are large by the standards of other molluscs. All living cephalopods are marine predators, and most lack an external shell. Familiar representatives include squids, octopuses, and cuttlefish. The only living genus with an external shell is *Nautilus*, found on reefs in the tropical western Pacific. It is the only representative of a very large number of extinct shell-bearing cephalopods. The shells of these animals are univalves differing from those of other molluscs in that the shell interior is divided into chambers. The partitions (septa) between the chambers

Posterior

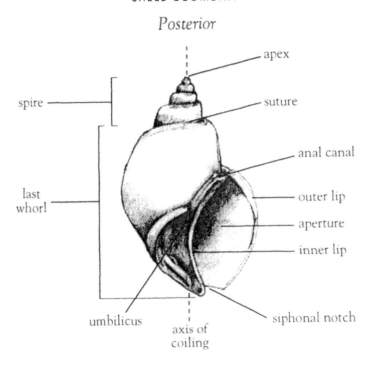

spire

last
whorl

apex

suture

anal canal

outer lip

aperture

inner lip

umbilicus

axis of
coiling

siphonal notch

Anterior

Fig. 2.2. Typical gastropod shell, showing parts of the shell frequently discussed
throughout the book.

are thin, mineralized membranes perforated by a posterior extension
(or siphuncle) of the main part of the animal's body, which occupies
the most recently formed part of the shell, or body chamber.

Clams constitute the second largest molluscan class, known as the
Bivalvia or Pelecypoda. The shell consists of two valves, one on the
right, the other on the left side of the animal (figs. 2.3, 2.4). Dorsally
they are joined by an elastic ligament, which acts as a spring to keep
the valves slightly apart while the animal is filtering water or ingesting
sediment for food. The shell is shut by the contraction of one or two
adductor muscles that connect the interior surfaces of the two valves.
Most clams are marine, but several lineages have successfully invaded
fresh water. All but a few are filter-feeders or mud-ingesters; some
deep-sea species, however, prey on small crustaceans by using specialized siphons to suck them up and trap them in the cavity between the
mantle and the gills.

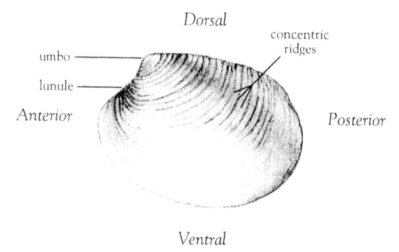

Fig. 2.3. Typical valve of bivalve shell, external view. The specimen illustrated is *Humilaria kennerleyi*, Friday Harbor, Washington.

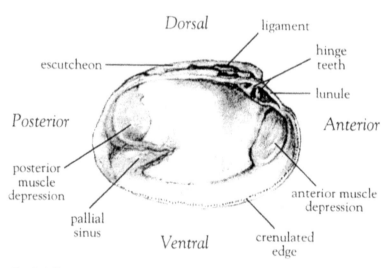

Fig. 2.4. Typical valve of a bivalve, internal view. The specimen shown is the same as in figure 2.3.

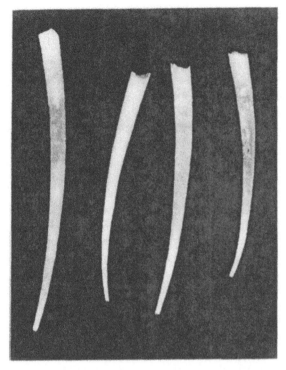

Fig. 2.5. Typical scaphopod, *Dentalium eboreum*, Sanibel Island, Florida. The shell is gently curved in planispiral fashion. The largest specimen illustrated is 30 mm long.

Members of the class Scaphopoda (tusk shells) have a gently curved tubular shell that is open at both the growing apertural end and the narrow apical end (fig. 2.5). They are predators of small animals in marine sands and muds.

The extinct class Rostroconchia is characterized by a univalved shell that superficially resembles the shell of a clam. Dorsally, the shell is somewhat flexible, so that the right and left halves could be manipulated to meet along a line of contact ventrally. Rostroconchs lived throughout the Paleozoic era (550–250 million years ago) in marine sandy and muddy habitats, and are believed to have had modes of life similar to those of living clams.

The Logarithmic Spiral and the Conservation
of Shape

Beneath the great diversity of forms there is a surprising unity of molluscan design. Molluscs build shells according to a few basic principles that relate growth and form. The most important of these is that, to a first approximation, the shell remains the same shape as it grows. If one could unwind a shell from its coiled state, the shell would be a hollow cone, closed at its small or apical end and open at its broad or apertural end. Growth of the shell takes place at the rim of the aperture and nowhere else. Each point on the rim can be traced back through successively earlier growth stages to the apex, which is the oldest part of the shell. As the shell grows, its aperture expands. The overall shape of the cone, however, remains the same.

Some shells look remarkably like simple cones, limpets being the best example (fig. 2.6). Close inspection, however, reveals that the cone is not really straight; its axis (the so-called translation axis, along which growth takes place) is a curve rather than a straight line. In the vast majority of snails, clams, and shell-bearing cephalopods, this curvature is so strong that the conical tube is a spirally coiled structure of one of more revolutions or whorls. The aperture revolves around an axis (the axis of coiling, or coiling axis) as it expands.

It was recognized by early naturalists that this curvature closely approximates the form of a particular kind of spiral, known as the logarithmic or equiangular spiral. In this curve (fig. 2.7), the distance between adjacent coils increases by a constant factor as one moves away from the center (or origin) of the spiral. This factor is one measure of curvature. If it is very large, the spiral expands rapidly away from the origin, and curvature is slight; if the distance between adjacent coils increases slowly, the spiral's curvature is great. Another way to express curvature of the spiral is by measuring the angle between a line drawn from the center of the spiral and crossing all the coils of the spiral, and a line that is tangent to the spiral (i.e., touching but not crossing the spiral at a given point). This angle is constant for a given logarithmic spiral (fig. 2.7). If the angle is 90°, the spiral collapses to a straight line. This happens when the radius and the tangent are one and the same line. At the opposite extreme, when the constant angle is a right angle (90°), the spiral becomes a circle.

The simplest kind of curved or coiled shell is a symmetrical one like *Nautilus* or the curved tube of a scaphopod (fig. 2.5). The left and right sides are mirror images of each other, separated by a plane of symmetry. In these so-called planispiral shells, the aperture can be

Fig. 2.6. Limpets. Top. *Patelloida chamorrorum*, Tarague Beach, Guam. The shell, which is 20 mm long, is a nearly perfect cone, with the apex close to the center. Bottom, *Lottia digitalis*, Piedras Blancas Point, California. The apex of the shell is displaced toward the front. Strong folds or ribs extending from the apex to the margin characterize the sculpture of this species. This specimen is 23 mm long.

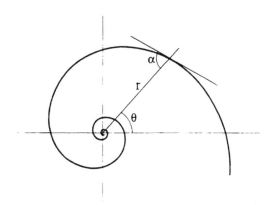

Fig. 2.7. Logarithmic spiral. The curve is illustrated in a polar coordinate system. The position of a point on the curve is determined by two quantities, r (distance from the point to the center, or origin, of the coordinate system) and θ (the angle between a radius and the horizontal line to the right of the origin). The general equation of the logarithmic spiral is $r = ae^{\theta \cot \alpha}$ where a is a constant (the radial distance from the origin of the coordinate system to the beginning of the spiral), α is the constant angle between the radius and a tangent to the curve, and e is the base of natural logarithms ($e = 2.72$).

Fig. 2.8. *Tectus triserialis*, Malakal Island, Palau. This trochid snail has the axis of coiling steeply inclined to the plane of the aperture, so that the shell's apex points up when the shell rests on a horizontal surface. The tubercles on the external surface are not covered by subsequent whorls as the shell grows. This snail is 43 mm high and 36 mm in diameter.

thought of as having a planar orbit around the axis of coiling, enlarging constantly as it makes successive revolutions.

If the aperture follows a three-dimensional trajectory as it revolves around the axis of coiling, the shell becomes asymmetrical and is then said to be conispiral (fig. 2.8). This is the situation in most snail shells and clam valves. The shell assumes a conical shape, the axis of the cone being the axis of coiling. As the aperture revolves and expands, it migrates along the axis away from the apex at a constant rate.

The Diversity of Shell Shapes

The simple cone, planispiral shell, and conispiral shell allow for progressively more variation in shape as the number of geometric variables needed to describe shape increases. For the simple cone, we need only three variables to specify form: the relative height of the cone, the shape of its opening, and the position of the apex. If the cone is tall and narrow, the rate of expansion of the aperture is low as the aperture moves along the translation axis away from the apex. If, on the other hand, the cone is low, the aperture's expansion rate is high. The shape of the opening may vary from almost circular, as in the tall North Pacific limpet *Acmaea mitra*, to narrowly elliptical, as in some limpets such as *Tectura paleacea* living on the narrow blades of sea grasses. The apex may be central in position, or be eccentric, lying toward the front or back end of the shell.

In planispiral shells, there is a fourth variable to consider. This is the degree of curvature. Together with the expansion rate of the aperture, curvature determines how tightly coiled the shell is. In loosely coiled shells, there is no physical contact between adjacent whorls. This usually means that the aperture expands relatively slowly and that curvature of the shell is slight. Tight coiling results from an increase in expansion rate of the aperture or from an increase in curvature or both. The whorls may be so tightly coiled that new whorls almost entirely envelop the earlier ones.

A full description of conispiral shells requires two additional variables. One of these is the height of the spire, that is, the angle at the apex of the cone. In a high-spired shell, such as that of an auger snail (Terebridae), the apical angle is small (as little as 3° in some needle-shaped species), and the aperture moves rapidly along the axis of coiling as it revolves and slowly expands during growth. Low-spired shells have a much higher apical angle.

The final variable is the inclination of the axis of coiling. When a snail crawls on a horizontal surface, the aperture faces down. In some shells, the axis of coiling is steeply inclined to the horizontal plane of the aperture, so that the apex points obliquely up (fig. 2.8). In other shells, in which the axis of coiling makes a very low angle with the horizontal, the spire and apex point mainly backward. In fact, the apex in these shells is usually the posteriormost part of the moving snail (figs. 2.9, 2.10).

There is an interesting relationship between the inclination of the axis of coiling in conispiral shells and the range of shapes of the aperture. In shells in which the spire is steeply inclined to the horizontal aperture, the latter is typically circular to broadly oval in outline. The apertural rim almost never contains a notch or channel at the front end of the aperture. In shells whose axis of coiling is oriented at a low angle to the horizontal, the range of apertural shapes is much greater. Besides shells with circular to elliptical apertures, there are many in which the aperture is long and narrow. Notches and other interruptions in the apertural rim are common features in such shells.

These differences in the diversity of apertural shapes make geometric sense. Revolution of the aperture about the axis of coiling can be thought of as having two components, one perpendicular to the plane of the aperture, the other within that plane. In shells whose axis of

Fig. 2.9. *Bursa tuberosissima*, Pujada Bay, Mindanao, Philippines. The axis of coiling makes a small angle with the plane of the aperture. The apex therefore points backward when the shell rests aperture down on a horizontal surface. The shell has complex varices (or thickenings) developed every 180° around the shell. As the shell grows, parts of these varices are covered by the next whorl and removed. An elaborate posterior or anal canal of the aperture is a characteristic feature. This specimen is 41 mm long.

Fig. 2.10. *Homalocantha anatomica*, Western Shoals, Guam. The adult shell is characterized by an extremely elaborate varix, which is much larger and more complex than previously formed varices. The animal thus displays determinate growth. Previously formed sculpture is partially obliterated through resorption as the inner lip encroaches on the outer surface of the last whorl to the left of the aperture. Left, apertural view; right, dorsal view. This specimen is 40 mm long.

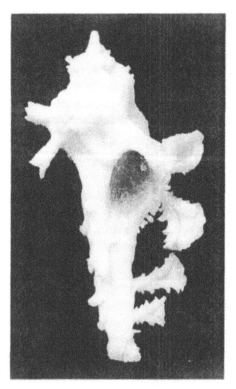

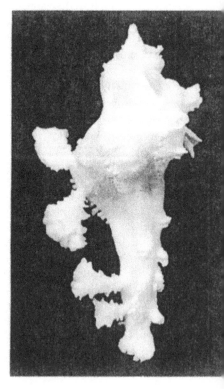

coiling is steeply inclined to the plane of the aperture, the component of rotation within the apertural plane predominates over the component perpendicular to the aperture. In the extreme and impossible case where the coiling axis is perpendicular to the plane of a rotating aperture, all points on the apertural rim would rotate around the axis and remain within the plane of the aperture. Under such absurd circumstances, the aperture would have to be perfectly round, and no interruptions in the rim could be maintained as the aperture rotated. These strictures are progressively relaxed as the axis of coiling becomes less steeply inclined to the aperture. In the extreme case where the axis is parallel to the apertural plane, all points on the apertural rim revolve around the axis perpendicular to the aperture. Interruptions along the rim can easily be maintained, and the aperture can take on a great range of shapes, because revolution does not interfere with the configuration of the rim.

Handedness and the Notion of Constraint

One of the first things most collectors of shells notice is that snail shells are usually coiled in a right-handed or dextral fashion. If the shell is held with the apex pointing up and the aperture opening toward the observer, the aperture appears on the observer's right. In left-handed (or sinistral) shells, the lip appears on the left (fig. 2.11).

The overwhelming predominance of right-handedness in snails raises some interesting questions. Is there something in the way shells are built that predisposes asymmetrical coiling to be dextral? Is right-handedness functionally superior to left-handedness? How does a reversal in coiling direction come about, and why are such reversals rare?

Answers to these questions have a significance well beyond the arcane world of shell geometry. It is a fundamental principle of evolutionary biology that many characteristics of organisms are conserved as lineages evolve. Were this not so, we would be unable to infer ancestor-descendant relationships, for these can be identified only by virtue of traits that the descendant inherits from its ancestor. The constellation of constant traits defines evolutionary branches or clades, and constitutes the so-called *Bauplan* (a German word meaning "building plan") of that clade. The conservation of traits means that evolutionary change in some directions is more difficult (or at least less likely) than in others. The design rules that define the *Bauplan* preclude variation in some traits while allowing variation and change in others. Evolutionary change is, in other words, constrained. The study of

CHAPTER 2

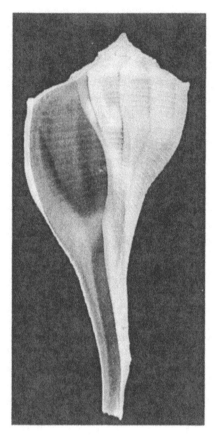

Fig. 2.11. *Sinistrofulgur sinistrum* (lightning whelk), Turkey Point, Florida. This left-handed species has a long spoutlike siphonal canal and numerous evenly spaced riblets on the inside of the outer lip. Although the species shown is 114 mm, the lightning whelk is one of the largest living snails, attaining a length of up to 40 cm.

handedness in snail shells provides some insights into what constraints are, how they operate, and how absolute they are.

It is evident from comparisons between snails and conispiral fossil cephalopods that right-handedness is not more likely to evolve than sinistral coiling. During the Silurian and Devonian periods of the Paleozoic era, five or six lineages of cephalopods independently evolved conispiral coiling. At least two or these had sinistral shells, and at least two were dextral. During the succeeding Mesozoic era, three lines became conispiral. The most diverse of these was the ammonoid family Turrilitidae, which in the early phase of its history during the Al-

22

bian stage of the Early Cretaceous showed coiling in either direction, even within the same species. Later, during the Cenomanian stage of the Late Cretaceous, species tended to be sinistral. There was, in other words, no predisposition toward one direction of asymmetry or the other among the eight or nine independently evolved conispiral clades of cephalopods. This observation strengthens the impression that the predominance of right-handed coiling in snails derives from some attribute that snails have and cephalopods lack.

One process that sets snails apart from all other molluscs is torsion. During the earliest phases of the development of individual snails, the body of the animal is conventionally organized as it is in other molluscs. There is an anterior head-foot complex connected by a narrow waist to the posterior part of the body containing the organs of digestion and reproduction. The cavity between the mantle skirt and the internal organs opens posteriorly. The process of torsion during the larval phase changes this plan of organization. As a result of the twisting of the narrow waist, the opening of the mantle cavity comes to lie over the animal's head rather than posteriorly. As seen from above, this twisting or torsion through 180° is counterclockwise in right-handed snails and clockwise in left-handed ones.

It is tempting to conclude that the prevalence of right-handed coiling in snails is linked to this unique process of torsion. However, torsion does not automatically result in conispiral coiling. Many adult limpets, for example, have symmetrical cap-shaped shells that are not conispiral, yet they undergo torsion as larvae. Moreover, the preferentially counterclockwise twisting of the waist between the head-foot region and the rest of the body is itself a symptom of an asymmetry that already exists in the fertilized egg of the snail. It is this asymmetry that predisposes snails toward, but does not inevitably result in, right-handed coiling.

Noel Morris at the Natural History Museum in London has pointed out that the earliest univalve shells of the Cambrian period, at the dawn of molluscan evolution, did not show a predominance of dextral coiling. Many of the minute univalves of the Early Cambrian had nearly planispiral shells in which the spire deviated slightly to either the right or the left. Morris believes that these molluscs, in which the direction of asymmetry was not rigorously specified as it is in living snails, were untorted. Conispiral coiling in some of the right-handed Cambrian forms became more accentuated, and it may have been that torsion arose in these animals. Torsion may have occurred not in the larval phase, as it does in snails today, but in crawling miniatures of the adult. The timing of torsion in the developmental sequence became progressively earlier, so that torsion was completed before the end of the larval phase. This earlier timing may have been accompanied by the establishment of regulations that specified the direction of asymmetry. Early left-handed groups never became as high-spired as did some of the right-handed univalves. It is thus possible that the functional benefits of torsion would have been greater for high-spired forms than for univalved molluscs that departed only a little from the planispiral condition.

Why did torsion not arise in cephalopods? Of course, we can never know the answer to such a hypothetical question, but from time to time it is instructive to think about why certain biological designs have not evolved. Cephalopods, in contrast to most snails, are swimmers rather than crawlers. It was long thought that torsion conferred an advantage to swimming snail larvae because it enabled the most vulnerable parts of the larva—supposedly the head and the velum used in swimming—to be retracted into the shell before the more resistant operculum-bearing foot. It turns out that most predators of larvae swallow the latter whole; in other words, the entire larva is at risk, not merely its head or its foot. In experiments using various predators of swimming larvae, Timothy Pennington and Fu-Shiang Chia at the Friday Harbor Laboratories showed that there was no difference in vulnerability between pre-torted veliger larvae and larvae of the same species that had undergone 90° of torsion. The main benefits of torsion may accrue to crawling juveniles rather than to swimming animals. Accordingly, torsion would be unlikely to evolve in a lineage of animals basically adapted as swimmers.

Morris's scenario could explain the origin of torsion in a right-handed ancestral univalved mollusc, but it is silent on the question of why transitions from dextral to sinistral coiling in torted snails are rare. The answer to this question, I believe, lies in the inferiority of the transition rather than in any disadvantage that left-handedness itself presents.

There are two ways in which coiling direction can be changed. The first is a simple reversal of the direction of torsion. The direction of torsion and thus of shell coiling is controlled by a single gene inherited through the mother. The gene comes in two forms (or alleles). In the freshwater lymnaeid snail *Radix peregra*, mothers with right-handed offspring have either two copies of the allele coding for right-handed coiling, or one copy of this allele and one for the allele resulting in sinistral coiling. Snails whose offspring are left-handed possess two copies of the left-handed allele. This means that right-handedness is genetically dominant over left-handedness in this species. In land

Plate 1 (apertural view); plate 2 (dorsal view). *Lambis (Millepes) scorpio*, off Babeldaob, Palau. The greatly expanded spinose outer lip of the adult provides lateral stability. Backwardly and forwardly directed spines further add to the shell's stability, and probably also provide protection against being swallowed by fishes. The outer lip has a curious series of short projections between the anterior siphonal canal and the notch through which one of the snail's eyes protrudes. Like other features of the adult lip and aperture, this series of projections is not well understood functionally. This specimen is 103 mm long including the spines.

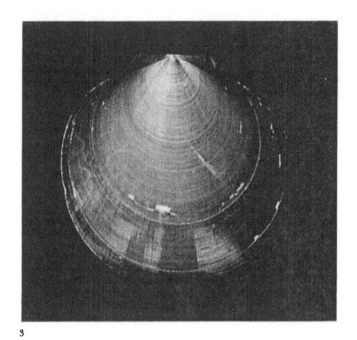

3

Plate 3 (upper valve exterior); plate 4 (lower valve exterior); plate 5 (lower valve interior). *Amussium japonicum*, Tosa Bay, Japan. This is a highly specialized swimmer, whose valves are flattened and smooth. The shell is internally strengthened by radial ribs, visible on the inner shell surface. There is no byssal notch. The specimen shown is 118 mm in diameter.

4

5

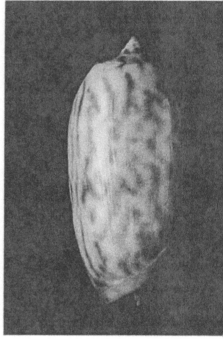

Plate 6 (apertural view); plate 7 (dorsal view). *Oliva minacea*, Majuro Atoll, Marshall Islands. The shell in life is usually enveloped by lobes of the large foot. By virtue of the large foot, this species can burrow rapidly in sand. The shell nonetheless provides good passive protection as well, because the foot can be fully withdrawn into the long narrow aperture. The whorls in the short spire are stepped, giving a ratchet effect that may prevent back slippage during burrowing. There is a well-developed notch at the posterior end of the aperture, where sense organs are located. This specimen is 52 mm long.

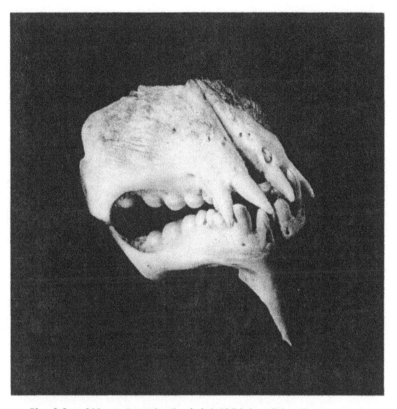

Plate 8. Jaw of *Monotaxis grandoculis*, a lethrinid fish from Palau. The front teeth are modified into grasping incisorlike devices, whereas the back teeth are molarlike crushing structures. Unlike the molars of mammals, which can move side to side and in a fore-aft direction, the molars of *M. grandoculis* and of other fishes can only crush.

9

Plate 9 (apertural view); plate 10 (dorsal view). *Phyllonotus regius* (Muricidae), Playa Brava, Costa Rica. The shell is adorned with complex varices. On its underside, there is a thick glaze or callus. This specimen is 116 mm long.

10

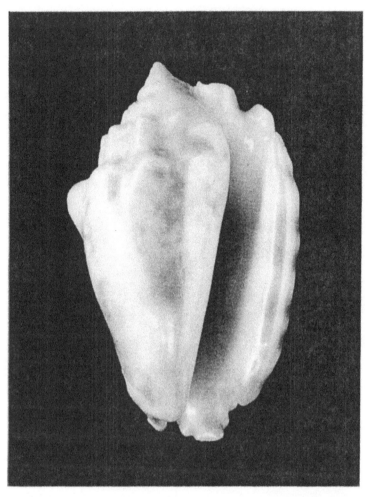

Plate 11. *Strombus (Lentigo) lentiginosus* (Strombidae), Ngemelis, Palau. The dorsal varix and glazed underside of this sand-dweller evolved quite independently of those of muricids and strombids. This specimen is 71 mm long.

Plate 12 (lateral view); plate 13 (apertural view). *Cypraea (Mauritia) mauritiana* (Cypraeidae), Pagan Island, northern Mariana Islands. This striking shell combines several notable defenses, including a smooth glossy surface, a narrow aperture lined by well-developed teeth, a greatly thickened and expanded base, and a tiny spire completely hidden from view by thick glaze. Both ends of the aperture are conspicuously notched. This specimen is 73 mm long.

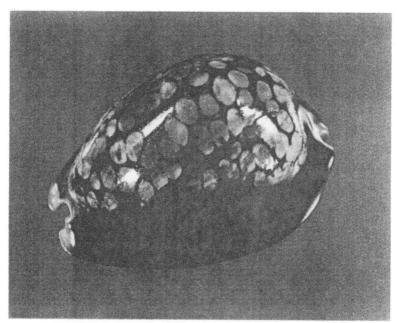

12

13

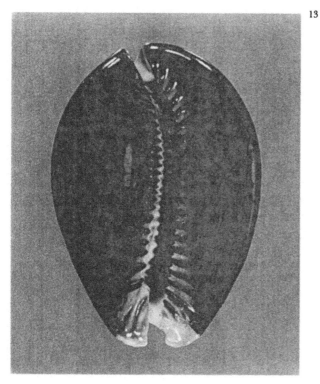

14

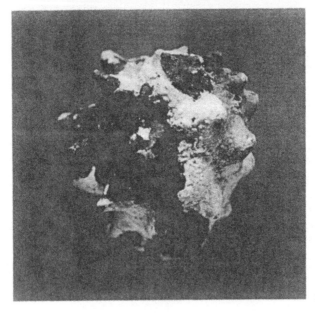

15

Plate 14 (apertural view); plate 15 (dorsal view). *Drupa morum* (Muricidae), Guguan, northern Mariana Islands. This extraordinarily heavily armored shell bears knobs on its dorsal surface. The aperture of the adult is bordered by a greatly thickened outer lip, which on its inner edge bears large teeth that partially occlude the opening. The broad ventral callus pad reflects a large foot, with which the snail clings to rocks on wave-exposed rocky shores. This specimen is 36 mm long.

Plate 16 (apertural views). *Terebra guttata* (110 mm long) (Terebridae), Majuro Atoll, Marshall Islands (right) and *T. areolate* (105 mm long), Luminau Reef, Guam (left, spotted). The orange shell of *T. guttata* is exceptionally high-spired. The high-spired spotted shell of *T. areolata* has a repair mark indicating an unsuccessful predation attempt by a calappid box crab.

Plate 17. *Terebra guttata* (right) and *T. areolate*, (left, spotted), enlarged views of dorsal sides.

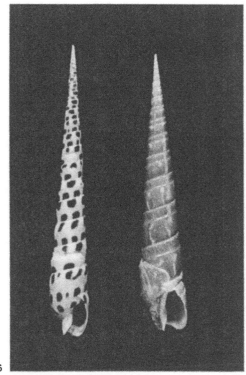

16

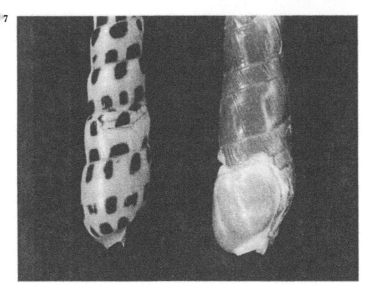

18

Plate 18 (interior view); plate 19 (exterior view). *Mexicardia procera* (Cardiidae), Playa Venado, Panama. Radial ribs extend beyond the valve margins anteriorly and form an effective grillwork even when the valves gape. This specimen is 70 mm high and 57 mm long.

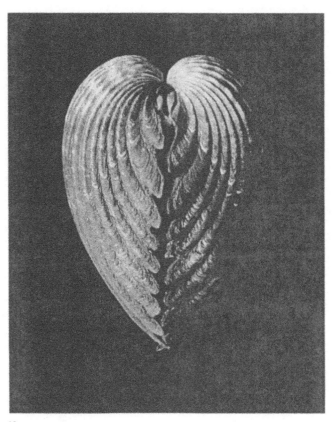

19

20

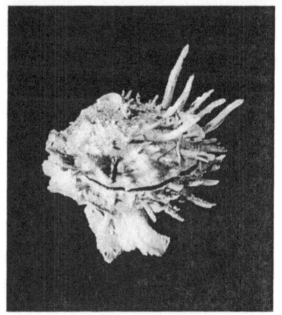

21

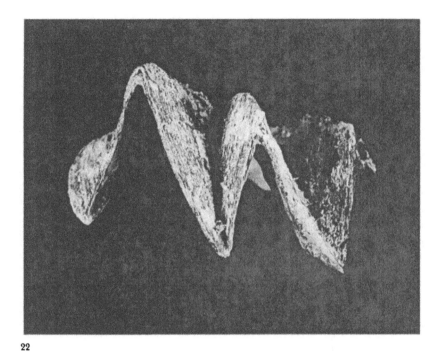

22

Plate 20 (gaping view); plate 21 (closed view). *Spondylus* sp., near Koror, Palau. Long spines extend beyond the valve margins, which are crenulated on their inner edges. All spondylids live attached to rocks. This specimen is 69 mm in height.

Plate 22. *Lopha crestagalli* (Ostreidae), near Koror, Palau. The valve margins are in the form of a zigzag. This species reaches a diameter of 12 cm.

snails of the genera *Partula* and *Laciniaria*, however, the left-handed allele is dominant over the dextral one.

Gary Freeman at the University of Texas at Austin has suggested that abnormally coiled individuals fail to produce a substance that is present in individuals with normal handedness. This failure may result in deleterious side effects. In the freshwater viviparid snail genus *Campeloma* and the land snail *Cerion*, rare sinistrals are not precise mirror images of their normal dextral counterparts. Stephen J. Gould and his colleagues at Harvard discovered that left-handed *Cerion* from the Bahamas have fatter shells than do dextral individuals. Moreover, the axis of coiling in the sinistral shells is slightly twisted, a condition not observed in the normal right-handed snails.

These findings may mean that some of the biochemical machinery that regulates growth and form in normal snails is defective in individuals with reversed coiling. The tentative conclusion therefore is that, although left-handed and right-handed shells may be functionally equivalent, the developmental disruptions associated with the reversal of coiling direction are often sufficient to prevent the establishment of mutations affecting handedness.

Yet, there are circumstances in which alleles coding for a reversed coiling direction become fixed in populations of snails. Three snail lineages, for example, have become left-handed in the North Pacific: one in the turrid genus *Antiplanes* and a second in the buccinid whelk genus *Pyrulofusus*, both beginning in the early Middle Miocene (about 16 million years ago), and the buccinid whelk *Neptunea leva* living in the Sea of Okhotsk. In the North Sea basin of western Europe, two lineages (whelks of the *N. contraria* group and the extinct terebrid auger shell *Terebrellina inversa*) became left-handed during the Pliocene, about 4.5 million years ago. At about the same time, a lineage of large cone shells (genus *Contraconus*) and the gigantic lightning whelks of the genus *Sinistrofulgur* became left-handed in Florida. The whelks are still living (fig. 2.11), but *Contraconus* alas is extinct.

What are these unusual circumstances that enable left-handed populations to become established? In a survey of reversals from right-handed to left-handed coiling that have occurred in marine snails during the Cenozoic era (the last 65 million years of earth history), I found that 12 of 13 reversals took place in lineages whose larvae emerge from eggs as crawling juveniles shaped like miniature adults, and that only one emerges as a swimming larva. The much more numerous reversals in land and freshwater snails also involve lineages lacking a swimming larval stage. I would interpret this to mean that the protective environment of the embryo growing inside a large food-rich egg is much more forgiving of alterations to the pattern of development than is the environment inside a small food-poor egg in which the embryo spends a short time before taking up life as a tiny swimmer. Those marine snail groups in which sinistral species have arisen all live or lived in sand or mud. Geographically speaking, seven lineages occur in cool-temperate or polar waters, and six are known from the tropical Atlantic. These environments, as well as those in fresh water and on land, may be less antagonistic to the establishment of oddities in shell geometry than are marine rocky bottoms or the many varied environments of the tropical Pacific and Indian Oceans, where no populations of left-handed species are known to have evolved during the Cenozoic.

The second way to bring about a change in coiling direction is by a remarkable process known as hyperstrophy. In a normal shell oriented with the axis of coiling vertical, the apex points up. As the aperture revolves around the axis, it also moves downward parallel to the axis away from the apex. In hyperstrophic snails, the aperture moves upward, so that the apex points down instead of up. An observer unaware of the fact that the shell is hyperstrophic would naturally orient the shell with the apex pointing up. A right-handed hyperstrophic shell would then appear to be left-handed. The true orientation is revealed by the soft parts of the snail, whose organization—loss of the right gill and right auricle, for example—is the same as that of a true right-handed snail.

The transition from a normal to a hyperstrophic shell implies an intermediate stage in which coiling becomes looser. We can see this by doing a thought experiment. If we take a normal right-handed shell and push the apex through the middle until it emerges at the opposite end, we will have created a hyperstrophic shell. This cannot be done without loosening the coils, that is, without reducing the overlap or contact between adjacent whorls. This loosely coiled transitional phase may be disadvantageous to many snails, because it compromises shell strength by making each whorl more vulnerable to breakage than it would be if there were greater overlap with adjacent coils. Hyperstrophy is nevertheless known to occur in the larval stages of most opisthobranchs (bubble shells and their relatives), architectonicids (sundial shells), and some freshwater apple snails from Africa (members of the genus *Lanistes* in the family Pilidae).

What does a change in coiling direction tell us about constraints? If a constraint is a circumstance preventing change in one evolutionary direction while channeling it in another, there can be little doubt that the rarity of left-handedness in snails represents the widespread existence of a constraint. The stricture is not, however, an absolute prohibition. Resistance to a change in coiling is clearly less in land and freshwater snails than in most marine ones. The constraint is therefore not a geometric or developmental one, but a resistance to change imposed by external circumstances. In other words, there is nothing in the developmental sequence or in the architecture of shells that prohibits a switch in handedness; instead, the transition may be functionally unacceptable in many environments. There may, in fact, be very few constraints imposed on form by design rules alone. Almost any *Bauplan*, no matter how rigidly determined it may seem to be, can be modified or even abandoned given the right external circumstances.

Growth and Form: The Inadequacy of the Logarithmic Spiral

Nature is rarely as simple as we might like it to be. The idea that shells are variations on the theme of the logarithmic spiral is so elegant in its simplicity and in its power to explain the diversity of shell shapes encountered in nature that we have trouble modifying it, much less letting it go. The inescapable problem is, however, that most shells change in shape as they grow, and therefore do not conform precisely to the logarithmic spiral. This change in shape might be interpreted as imprecision on the part of molluscs; if external conditions did not interfere with orderly shell deposition, then the mollusc could build the shell "correctly" in accordance with the logarithmic-spiral model. Another way of looking at these deviations, however, is to think of them as normal. They may be the expression of additional principles of form that apply to all shells and that, in the end, show our conception of shells as logarithmic spirals to be inadequate and incomplete.

One of the missing principles is the relationship between growth and form. The shape of a shell is determined by the way skeletal material is added by the mantle to the apertural rim. When the shell is being enlarged in the spiral direction, the mantle margin lies at the growing edge. Not all the layers of the shell are deposited at once. The first to be laid down is the outer layer or periostracum, which is made of protein and usually lacks a mineral component. Then comes a mineralized layer, which provides rigidity to the framework of the periostracum. Finally, the inner layer of the shell is added. This occurs well back from the shell edge, and is accomplished by the mantle's surface rather than by its margin. When growth in the spiral direction is not taking place, the mantle margin withdraws from the edge of the shell, and only the innermost shell layer continues to be deposited.

Growth is not a continuous process. It takes place at intervals separated by times of quiescence. The episodic nature of growth is indicated on the shell by the presence of growth lines. These represent previous positions of the apertural rim and, despite their name, record times when growth temporarily ceases.

When a shell grows by extending its aperture in a spiral direction, each point on the shell edge moves a short distance to a new position. This movement has two components, magnitude (or distance between the old and new positions) and direction. In addition, the aperture is expanding as the shell grows. Any two points on the apertural rim at the new position lie a little farther apart than they did in the old position. This is why adjacent spiral ribs on snail shells and radial ribs on clam valves diverge from one another as the shells enlarge. Secondary ribs often appear in the gaps between the major ribs as the latter diverge. The proliferation of secretory cells at the mantle edge is responsible for this expansion.

The rate of growth is extremely sensitive to environmental conditions and also varies according to an individual's age. In most molluscs, growth is faster at high temperatures and in the presence of abundant food than in cold nutrient-poor conditions. Temperate and polar species thus typically grow fastest in summer and little if at all in winter. During an individual's life span, growth is typically fastest early in life. After reaching a peak, it tends to fall off as the animal approaches maturity. Growth ceases altogether in species with determinate growth. Slow growth is usually indicated when growth lines are crowded together and when the number of growth lines produced in a year is small. Growing slowly therefore means not only that the distance traversed by points during growth intervals is short, but also that these intervals are infrequent.

In a shell that grows precisely according to the logarithmic spiral, the direction of growth is always the same regardless of the distance traversed by a point during an interval of growth. In other words, growth direction is independent of growth rate. This independence, however, probably never occurs. In fact, I suspect that the deviations from logarithmic-spiral architecture are caused by basic relationships among growth rate, growth direction, and apertural expansion.

It is becoming clear from observations and experiments that rapid growth is associated with high rates of expansion of the aperture. This relationship was experimentally verified by Paul Kemp and Mark D. Bertness at Brown University. They maintained two groups of periwinkles (*Littorina littorea*), both drawn from the same population, under contrasting conditions in the laboratory. One group had available a plentiful supply of seaweeds as food, whereas the other was maintained under crowded conditions with much less food available. Snails in the first group grew faster and developed low-spired shells with a large body whorl and a rapidly expanding aperture, whereas those in the second group grew more slowly and had higher-spired shells with a smaller opening. Similar transformations are obtained when limpets are transplanted to rocks with different amounts of food present. When limpets are moved to rocks thickly covered with seaweed, shell growth speeds up and the aperture flares out as it expands at an increasing rate. A slowdown of growth has the opposite effect of reducing apertural expansion, with the result that the shell becomes a relatively taller cone whose sides are convex outward. In coiled snails, changes in shape are also evident as growth rate changes. Early in life, when growth speeds up and the aperture expands rapidly, the spire has a concave profile when seen from the side (fig. 2.12). As growth slows with age after reaching a peak, the aperture expands less rapidly and the spire comes to have a convex profile. In clams, the same principle yields valves that are increasingly inflated as growth slows down with age. In other words, valves of older and larger individuals are more convex than those of younger, smaller, faster-growing clams.

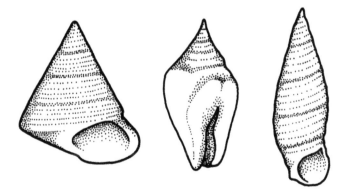

Fig. 2.12. Shells depart in various ways from strict logarithmic spiral growth, which is here represented by the shell on the left, in which the profile of the spire, as seen from the side, is straight. The most common deviation is illustrated by the shell in the center, in which the spire's profile is convex. The shell on the right represents a less common deviation, in which the profile is concave.

We do not know what mechanisms account for this interdependence of growth rate and apertural expansion. It is possible that, when the rate of cell division at the mantle edge declines, there is less outward pressure on cells and therefore a reduction in the rate at which the mantle edge expands.

The relationship between growth rate and expansion also applies to growth direction of individual points along the apertural rim. Some points move farther (grow faster) during a given interval of time than do other points because they lie at a greater distance from the apex. Put another way, the length of the spiral connecting a point on the rim with the apex is longer for some points than for others, yet all points on the rim traversed these different distances in the same amount of time. This relationship is perhaps best appreciated in clams. In the mussel *Mytilus edulis*, for example, points on the shell edge lying close to the umbo (or apex) grow very slowly, whereas those lying on the posterior edge opposite the umbo grow most rapidly. Points at fast-growing parts of the edge move away from the axis of coiling at faster rate than do points in slow-growing sectors.

This conception of how molluscan shells grow makes us see shells in a different architectural light. When the logarithmic spiral is seen as the ideal model for shell architecture, departures from it are considered aberrations that require special explanation, often in functional or adaptive terms. In the view I advocate here, however, deviations are expected because of the geometric relationship between growth rate and growth direction and expansion. This relationship modifies shells that are built on a logarithmic-spiral plan of architecture.

Some deviations cannot be explained simply by evoking the principle that expansion rate declines as growth slows. For example, some snails, such as many species of *Conus* (Conidae) and olive shells (Olividae), have a markedly concave profile of the spire when seen from the side (fig. 2.12). It turns out that this type of deviation is common in snails in which the posterior end of the aperture is marked by a notch or canal. In *Terebellum* (Seraphsidae) and in olive shells, this notch is occupied by a sense organ that detects danger from behind while the animal is buried in the sand. Detection requires that the sense organ remain at the shell margin. Posterior extension of the aperture may thus continue even if growth along other sectors of the apertural rim is not taking place. This would thus result in a spire with a concave profile.

Other variations result from the ability of molluscs to modify the pattern of episodic growth and to remodel the shell interior. With these elaborations, the *Bauplan* of the molluscan shell allows for a remarkable range of architectural possibilities and can be made compatible with almost any mode of life.

Episodic Growth

In the shells we have discussed so far, the aperture expands gradually and consistently as the shell grows. A very widespread departure from this condition occurs in shells that bear elements of sculpture produced at more or less regular intervals as the shell grows. If these elements are parallel to the growing margin, they are said to be concentric, axial, or collabral (parallel to the lip) (fig. 2.3). Others may be oriented obliquely to the shell edge. Still other elements are beads, knobs, tubercles, or spines representing the periodic elaboration of spiral or radial ribs oriented perpendicular to the growing margin. Each time such a periodic feature is formed at the apertural rim, the opening or part of it expands and changes shape slightly. The aperture thus expands in stepwise fashion and varies in shape according to whether its rim is a periodic elaboration or in a stage between elaborations.

Such changes in apertural size and form may be quite profound. The shells of many muricid, ranellid, bursid, and cerithiid snails, for example, produce thickenings (varices) at regular angular intervals during growth (figs. 2.9, 2.10). In many tropical members of these families and in the North Pacific muricid *Ceratostoma*, varices appear at intervals of 90°. In frog shells (Bursidae) and some muricids such as *Eupleura* and *Aspella*, as well as in the ranellids *Gyrineum* and *Biplex*, varices are separated by 180°, with the result that varices on adjacent whorls line up one behind the other. The impression is of a shell that has been squashed flat from above. In the personid genus *Distorsio*, varix formation is accompanied by a reorientation of the aperture. The coiling axis is parallel to the plane of the aperture when the apertural rim is a varix, but is tilted 20° above the plane when the apertural lip is simple. Varices in some conchs (Strombidae), nutmeg shells (Cancellariidae), and horn shells (Potamididae) are expressed mainly on the shell interior rather than externally.

Growth in the spiral direction in varix-bearing shells is intermittent but fast. When a growth spurt begins, a thin flexible extension is built beyond the varix. This layer is then mineralized and gradually thickened from within while a varix forms at the new outer lip. For a period of several days, the new addition is a very delicate part of the shell. While it is being formed, the builder tends to remain inactive and to be concealed under stones or in the sand. Once the new varix has been stiffened, growth in the spiral direction ceases. As the snail ages, intervals during which no spiral growth takes place increase in duration. Because varices are situated at approximately constant angular intervals around the shell, a greater distance must be covered between adjacent varices during later growth spurts.

A form of episodic growth unique to cephalopods is the formation of septa in the shell interior. From time to time the main part of the body extends forward in the shell. As it does so, the posterior surface of the mantle secretes a new septum that sets off a chamber. *Nautilus* typically has 30 or more such chambers and septa.

In many molluscs, growth in the spiral direction ceases when sexual maturity is reached. Shells displaying this determinate growth may be distinguished by the unique and often remarkable elaboration of the adult apertural rim. The adult lip of conchs (Strombidae) and many murexes is usually a broadly flaring structure, sometimes bearing spines that extend in various directions (fig. 2.10). This final lip is larger and more more complex than are the internal varices produced episodically during earlier growth stages. Cowries (Cypraeidae, Ovulidae, and their relatives) have a strongly in-rolled adult outer lip bearing denticles on its inner edge. In many land snails, as well as a few clams, determinate growth is indicated by the simple outward flaring of the outer lip (fig. 2.13).

Episodic and determinate growth can be thought of as exaggerations of the discontinuous growth pattern that characterizes all molluscs. As I pointed out in the preceding section, growth always proceeds in steps, as indicated by the presence of growth lines. In molluscs growing with periodic elaborations, the intervals of quiescence are lengthened, and individual growth steps may be greater. In some muricids and ranellids, in which there is a pattern of large varices interspaced with smaller axial nodes and very fine beads on the entire shell surface, there is probably a hierarchy of steps from very long (represented by the varices) to very short (represented by the beads).

Internal Remodeling

In the discussion thus far, I have simplistically assumed that all the shell material that is deposited is faithfully retained as the shell grows. The shell would then be a complete archive of the builder's life and times. Probably all molluscs, however, are capable of removing at least some of this mineral after it has been incorporated into the shell. Snails, for example, remove a thin layer of shell from the left border

Fig. 2.13. Growth series of the land snail *Cerion uva*, Boca Table, Curaçao. There is a dramatic change in shape as the shell grows. The young animal carries the shell with the apex pointing up, whereas the beehive-shaped adult, which is characterized by an expanded outer lip, crawls with the apex pointing backward. The animals are usually found at rest, clinging to plants or rocks by means of a film of dry mucus secreted by the foot. *Cerion uva* can reach a length of 24 mm.

of the aperture where the outer surface of the body whorl is about to be glazed over by the inner lip as the aperture expands. In muricids and many other gastropods, this resorption can be very extensive, resulting in the wholesale removal of spines and other sculptural elements (figs. 2.9, 2.10). This kind of sculptural remodeling is architecturally very useful. Varices, spines, knobs, and other forms of sculpture can get in the way of the inner lip of the aperture as it encroaches. As long as they can be resorbed, these features may be constructed anywhere on the body-whorl exterior and provide benefits related to strength, shell stability, and so on (see chaps. 4 and 6). In snails that cannot resorb large features but in which external sculpture is nevertheless useful, high-relief elements must be confined to those parts of the body whorl that are not enveloped by the next whorl. Remodeling therefore eliminates restrictions on where sculptural features can be placed on the external surfaces of tightly coiled shells.

Some snails change the configuration of the shell interior. In cone shells (Conidae), for example, about 25% of the shell material that is originally deposited is removed from the inner whorl partitions, which therefore become extremely thin. Externally, the shell is a thick-walled fortress with a long narrow opening. Without internal remodeling, there would be precious little room for the snail's soft parts, to say nothing of the prey worms and even small fish that the cone snail swallows whole. Thinning of the inner whorl partitions ensures, however, that the shell interior is quite capacious. In snails of the intertidal families Neritidae and Ellobiidae and the terrestrial family Helicinidae, so much shell material is removed from the internal surfaces that the spiral configuration of the shell interior has given way to a single vaulted chamber (fig. 2.14).

Another way in which resorption changes shell shape is by the loss of apical whorls. Resorption of the inner shell walls in the early parts of high-spired snail shells makes the shell so thin in this region that the apical whorls drop off. The remainder of the shell is then plugged by an apical wall or pad. This so-called decollation occurs widely in snails. It also characterized Paleozoic ascocerid cephalopods. In all these cases, the adult shell has a truncated, often cylindrical appearance.

Decollation has also recently been described in the northeastern

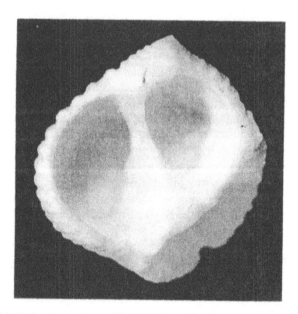

Fig. 2.14. *Nerita plicata*, Guam. The left side of the shell has been removed in order to show the shell interior. Instead of being spiral in form, as is typical of most other snails, the shell interior of the Neritidae is greatly modified by resorption of inner shell walls. A single partition walls off an apical chamber from the rest of the shell's interior. The specimen shown is 23 mm high.

Pacific tusk-shell *Rhabdus rectius*. From time to time, a part of the posterior end of the shell drops off as the shell wall is thinned from within. The removed part of the shell, which may be up to 5 mm in length, is then partially replaced by a secondary shell tube. The result is that the posterior aperture, through which the animal's siphons protrude, can enlarge as the shell grows in an anterior direction. Decollation and secondary shell formation are evidently episodic processes.

Arthropods and Molluscs Compared

The molluscan shell represents one very widespread way of building an external skeleton. Similar structures are found in brachiopods (lamp shells)—bivalved filter-feeders superficially resembling clams—as well as in the external armor of some vertebrates. A completely different form of exoskeleton, however, characterizes jointed animals or arthropods, the largest and by all accounts the most successful animal phylum. In arthropods, the skeleton is a jointed machine that is periodically shed and replaced by a larger version as the animal grows. Shedding (or molting) is an episodic process that is under hormonal control.

An obvious advantage of skeletal replacement in the arthropod mode is that large-scale modifications can be made to the skeleton as the animal grows and its functional demands change. Features useful to the small juvenile can be eliminated in successive molts, and innovations more useful to a larger animal can be introduced. These changes can occur anywhere in the skeleton, not merely at one end as in molluscan shells. Modifications are made on the architectural plan of the preceding molt, but the potential exists for far-reaching alterations in design as size increases.

One difficulty with molting is that the animal is nearly helpless between the time the old exoskeleton is shed and the new edition hardens. Not only is the body soft and unprotected against potential enemies, but the animal is incapable of moving, feeding, or defending itself. Consequently, molting generally takes place in secluded sites where the animal is in little danger. Small animals have little difficulty finding such shelters, but for large arthropods there are few such places. These potential problems do not apply to most molluscs, because the latter maintain their skeletons.

The molting pattern of arthropods closely parallels that of molluscs whose growth is episodic. In both groups, times of skeletal enlargement are often spent in hiding while the animal is inactive. The concentration of growth during particular intervals may be beneficial.

Growth tends to divert energy from other life functions; overall performance may therefore improve if growth is separated in time from most other functions. If, for example, all available resources are temporarily applied to growth alone, the latter might be much faster than it would be if resources were divided among several concurrent activities.

The Evolution of Geometric Diversity

The principal aim of this chapter has been to describe the geometric and developmental principles that dictate shell shape. In the course of evolution, these principles have themselves been elaborated, modified, and occasionally abandoned altogether.

Two apparently conflicting tendencies may be discerned in the evolutionary history of the rules of shell construction. The first is the tendency of increasing specificity or regulation. Aspects of form during early evolutionary stages are often free to vary because the regulatory machinery that specifies particular values or relationships is not yet well established. The second tendency is for the rules to be gradually extended, so that a greater variety of precisely specified shapes can be produced in the regulatory framework of the *Bauplan.*

The evolution of asymmetry in molluscs exemplifies the first tendency well. Earlier in this chapter I mentioned Noel Morris's hypothesis that slight deviations from planispiral coiling in early univalved molluscs were either to the right or to the left. The direction of asymmetry was probably not precisely regulated, because right-handed and left-handed shells appear to be very similar to one another and perhaps to belong to one and the same species. Only later, when torsion and its cascade of effects appeared at progressively earlier stages of snail development, did the direction of asymmetry become consistent.

We see the same phenomenon in conispiral cephalopods. In early members of the Turrilitidae, dextral and sinistral coiling appeared in equal frequency within populations. Later species tended to be only sinistral.

There is some intriguing evidence that the precision with which form is specified varies according to the habitat occupied by molluscs. Arthur Clarke at the U.S. National Museum of Natural History and Michael Rex at the University of Massachusetts at Boston have shown that snails and clams living in the cold waters of the Arctic and the deep sea are more variable in shape than are related species from warmer waters. Cold-water shells are, on the whole, more sloppily put together than are tropical ones; episodically elaborated sculpture is less regularly arranged, and subtle aspects of shell shape such as the outline of the valve margin and the position of the umbo in clams are not usually diagnostic of species, as they often are in warm-water molluscs.

One explanation for this pattern is that strict developmental controls on form may be important only in environments where large numbers of species coexist and where competition for resources is intense. Imprecision may be of little consequence if the risks of being substandard or different are low, as they may be when individuals live in circumstances where they encounter few enemies. If this interpretation is correct, it would mean that the environments in which the earliest molluscs evolved were less hazardous to individuals (i.e., more forgiving of small architectural transgressions) than are the environments of most living species.

Similar explanations may account for the fact that a switch from right-handed to left-handed coiling occurs mainly in snails in which the early developmental stages are passed in the safe confines of the mother or the egg instead of as a larva in the open sea. Embryologists have known for years that the pattern of animal development in large yolky eggs or in the body of the mother, where food is plentiful and threats from enemies are low, is much less conservative than is the often highly stereotyped pattern in embryos that come into the world at much earlier life stages. With the risks and hazards of early development reduced while in protective custody, even the most entrenched rules of construction can be changed. In other words, constraint—the precision and enforcement of rules of construction—is not absolute, but is influenced by the ecological consequences of deviating from the established order. If the cost of such deviation is low, transitions from one architectural plan to another become more likely.

Set against this increasing precision in the specification of form is the tendency for the range of well-regulated shapes to increase over time within a *Bauplan*. I have already discussed the increasing number of possibilities from the simple cone to the planispiral form to the conispiral configuration. Although all these architectures were established during the earliest Cambrian phase of molluscan evolution, it is likely that a simple straight or weakly curved cone was the ancestral condition in conchiferans, and that the conispiral form was derived several times independently from ancestors with planispiral coiling. Within the conispiral plan, the greatest diversity of apertural shapes occurs in shells in which the axis of coiling is tilted at a low angle above the plane of the aperture. This geometry appears to have been derived several times from conispiral ancestors in which the coiling axis was steeply inclined to the apertural plane and in which the potential diversity of apertural shapes is limited.

The earliest molluscs were already characterized by two additional growth characteristics, discontinuous growth and the ability to resorb deposited shell material. Elaboration and exaggeration of these characteristics in later lineages further expanded the range of evolutionarily attainable shell designs available to animals whose skeleton enlarges at one end. With the repeated evolution of determinate growth, far-reaching elaborations of the last whorl and apertural rim became possible. Evolutionary studies indicate that molluscs are primitively indeterminate growers, and that gastropod groups with determinate growth became prominent only in the last 20–30 million years or so. Snails capable of extensive internal remodeling arose during the Devonian period, about 400 million years ago, but again the groups that are capable of removing internal partitions and large elements of external sculpture have been prominent only during the most recent phases of molluscan history.

Shell geometry has attracted the attention of scholars for more than 200 years, yet we still know little about how the rules of construction work at the physiological level and how evolutionary change in form is brought about. The specification of rules in mathematical terms is a far cry from knowing how rates and directions of growth are determined at the mantle edge. We cannot even be sure that our abstract conception of form and of the rules governing shape is a good guide to the biological processes controlling growth. A major challenge for scholars of the future is therefore to seek a translation of abstract rules into the language understood by cells.

Meanwhile, we must make do with what we have, namely, an intuitive understanding of the geometric diversity of molluscan shells. We must now proceed to the question of how much of this potential diversity has actually evolved. This question lies at the foundation for the remainder of the book.

The Economics of Construction and Maintenance

In the preceding two chapters, I have approached the subject of shell construction and diversity from a geometric point of view. Building houses, however, involves more than drawing up architectural plans. The plans must be executed with the raw materials, labor, and capital that are available. It is therefore time to turn to the economics of shell construction.

In any discussion of form and function in biology, an essential component is cost. Certain designs may be extremely advantageous in survival or in reproduction, but if the costs of construction and maintenance are sufficiently high, such designs are economically and evolutionarily unattainable. Just as it is a major task of economists and entrepreneurs to identify the costs of doing business, it is important for evolutionary biologists to examine the costs in material, energy, and time required to build and maintain biological structures. We must uncover the ways by which these costs have been reduced and ask to what extent and in which circumstances these cost-cutting measures have been implemented by organisms.

How are such costs to be measured? In an absolute sense, the currency of the economy of life is a combination of energy and time, but just as the value of money depends on what can be bought with it, so time and energy become important only in the context of benefits that investment brings. It is the marketplace in which organisms compete for resources—food, mates, and shelter—that dictates costs and benefits. Predators and competitors as well as the nonbiological environment pose risks to survival and to reproductive success. The magnitude of these risks and of the potential benefits determines the value of the currency. For organisms, cost is failure or potential failure, whereas benefit is success as manifested by survival and reproduction.

In the case of molluscan shells, costs may be incurred in several ways. In the first place, the building materials—the mineral calcium carbonate and an organic protein matrix—may be limited in availability. In the second place, the energy and time required to emplace these materials into a functional shell may compete with the energy and time devoted to other life functions. Thirdly, even after shell material has been deposited, resources must be devoted to prevent the shell from deteriorating. This requires repair and maintenance, the ability to protect the shell from the everyday forces of weathering and use as well as from the more catastrophic events that cause large-scale damage. Finally, the presence of the shell exacts a burden. For molluscs capable of locomotion, shell transport requires energy. In short, fundamental incompatibilities can arise that must be resolved according to how survival and reproduction are affected.

This chapter is devoted to what little we know about the costs of constructing and maintaining a shell. I examine not only the energetics of calcification, but also some of the functional incongruities among strength, growth rate, locomotor performance, resistance to abrasion, potential for remodeling, and still other attributes. Many of the principles that emerge apply also to other groups of organisms, and are therefore important to an understanding of how economics influence evolution.

Mineral Availability

The water bath in which most molluscs live is rich in mineral salts appropriate as raw materials for skeletons. Whereas some organisms—diatoms, some sponges, and tintinnid protists, for example—have opted for silica (silicon dioxide), many others, including the molluscs, use calcium carbonate as the chief mineral component of their skeletons. Seawater in most of the world's oceans is supersaturated with calcium and dissolved carbon dioxide, so that availability of raw materials is not a significant limitation in the design of shells. Even in fresh waters, where salt content is generally lower, there is often enough calcium carbonate available that molluscs are capable of building massive shells, especially in areas of limestone. The same applies to many habitats on land.

In fact, it has been suggested by some that molluscs and other animals were faced not with a shortage of minerals, but with an excess that had to be purged from the cells at the time skeletons first arose during the Early Cambrian. Normally, animal cells contain calcium ions at a low concentration of about 10^{-7} M. At these safe levels, calcium performs a wide variety of critically important physiological functions, including transport and communication within cells. At higher concentrations, however, calcium becomes toxic. Before the

Cambrian, so the argument goes, calcium concentrations in the oceans were generally low. As the concentration rose, however, prevention of a calcium overdose required cells to excrete the excess in the form of a relatively insoluble mineral salt into which the calcium ions were incorporated. Proponents of this argument point to the widespread evolution of calcareous skeletons by animals, plants, and single-celled organisms near the beginning of the Cambrian as evidence that calcium concentrations in the ocean had reached a dangerously high level, and that the skeletons had therefore evolved as a way of solving a physiological waste-disposal problem.

Appealing as this argument might be, I believe it is wrong. Extensive deposits of limestone and dolomite, both of which contain large amounts of calcium, are known from well before the Cambrian, and therefore indicate very high concentrations of the offending calcium ions long before shelly fossils appeared in the record of life. Moreover, other minerals, notably silica, were adopted as skeletal materials during the Early Cambrian at the same time that calcium carbonate and calcium phosphate were. Still other animals developed organic-walled skeletons, and there must have been many animals that never developed a resistant outer cover at all, just as there are many such organisms in calcium-rich environments today. In other words, excretion of calcium ions in the form of a skeleton was not requisite for all organisms in calcium-rich habitats, and the fact that materials other than calcium carbonate were employed for skeletons suggests that there were sound functional reasons to build skeletons.

Even if the toxicity hypothesis were correct, the important point remains that the availability of raw materials for building the shell is not a problem for most molluscs. The widespread adoption of calcium-based salts as skeletal materials by organisms from Cambrian time onward suggests to me that supply has not generally been a limiting factor for molluscs at any time over the last 550 million years.

Although molluscs in most environments have access to a large supply of raw materials for building a shell, this is not so for all. Cold seawater, deep waters, and many fresh waters are undersaturated with calcium carbonate, so that molluscs living there may be affected by a limitation in mineral supplies. Calcium carbonate is unusual among mineral salts for being more soluble at low than at high temperatures in seawater. The exact relationship between solubility and temperature is highly dependent on the presence of other ions, but the biological ramifications are clear. First, molluscs living in calcium-poor cold waters must overcome a higher activation energy, or energetic hurdle, to precipitate calcium carbonate than do molluscs in warmer waters. There is, in other words, a higher energy cost associated with calcification in cold water. Secondly, the higher solubility of calcium carbonate in cold water means that shells there are subject to more rapid deterioration.

Energetics and Calcification

Like any other biological function, shell-building requires energy. Because the energy that is available to individuals is limited, a substantial energetic cost of constructing a shell would compete with other life functions. Some of this cost is incurred by the precipitation of calcium carbonate, and the rest arises from the synthesis of protein components of the shell. We are still far from understanding the energetics of shell construction fully, but the work of A. Richard Palmer at the University of Alberta is revealing some important principles and some estimates of energy cost. In experiments with the northeastern Pacific muricid gastropod *Nucella lamellosa*, Palmer estimates that about 1.6 J of energy are required to lay down 1 mg of calcium carbonate. This cost is about 5% of the energetic cost involved in laying down the protein matrix, which by weight accounts for not more than 2% of the shell of *N. lamellosa*. The 29.1 J of energy that a snail devotes to laying down the protein component of 1 mg of shell accounts for 10%–60% of the energy devoted to growth of the nonmineralized tissues of the animal, and for 15%–150% of the energy devoted to the production of eggs and sperm. These rough calculations therefore indicate that the organic component is energetically expensive, whereas the energy required for the precipitation of the mineral component is low. Palmer estimates that the organic matrix accounts for 22% of the energy required to build the shell if the latter is 1.5% organic by weight, whereas for a shell that is 5% organic, half the energetic cost of building a shell is due to the production of organic matrix.

It is clear from these findings that construction costs could be reduced substantially by decreasing the protein fraction of shells. Evidence that such a decrease is potentially meaningful in nature comes from Palmer's studies of shell repair in gastropods. Shells that were artificially damaged at the outer lip were subsequently allowed to regrow in the absence of food for a period of a little more than 1 month. The species with the lowest rates of repair had a nacreous or mother-of-pearl inner shell layer, in which aragonitic tablet-shaped crystals are arranged in layers or columns with organic matrix in between. In the three gastropods with nacreous structure that Palmer investigated, the organic content of shells varied from 4.1% to 4.4%. Repair in various other gastropods with a different microstructure and an organic content of 3.1% or less was much faster.

Are construction costs significant in the overall energy budgets of individual molluscs? The tentative answer to this question seems to be yes, at least in those few gastropods that have been investigated thus far. Palmer's work on *N. lamellosa* indicates that the rate of growth of the shell in the spiral direction sets an upper limit on the rate of soft-tissue growth. Under identical experimental conditions, thick-shelled and thin-shelled forms of *N. lamellosa* added the same mass of shell material over the duration of the experiment, but growth in the spiral direction was greater in the thin-shelled snails. In his work with the related *N. emarginata* in northern California, Jonathan Geller demonstrated that thin-shelled females were able to produce more offspring than were thick-shelled ones. Taken together, these results imply two things. First, there exists an upper limit to the rate of calcification and second, that snails can exercise some control over whether shell production occurs in the spiral direction, resulting in shell enlargement, or in thickening the shell wall. The rate of tissue growth and the allocation of resources to reproduction may also depend on the rate at which the shell grows.

The economics of calcification could in principle dictate the range of shell shapes evolutionarily available to molluscs. If costs of construction were high, only those shells in which a maximum volume is enclosed with a minimal amount of mineral might be feasible. Such economy could be achieved in at least two ways. First, the shell could be thin. Second, the shell could approximate a sphere. The sphere is the geometric solid with the lowest ratio of surface to volume for any given volume, and therefore the most efficient enclosure of space. The more a shell departs from the spherical form, the more material of a given thickness will be required to enclose a specified volume.

In univalved molluscs, conical and spiral architecture prevents shells from closely approximating a spherical shape, but tight coiling coupled with a moderately high expansion rate and a circular to broadly ovate aperture provides a relatively efficient way to enclose space, especially because parts of the shell that served as the outside wall become the inner partitions as the animal grows. If these partitions are secondarily removed by resorption, a quite reasonable approximation to a sphere can be achieved. Some species of the tropical high intertidal gastropod genus *Nerita* have exactly this architecture. At the opposite extreme among gastropods are shells with a high spire, low expansion rate, low whorl overlap, and a long narrow aperture.

Among bivalves, the spherical shape can be closely approached, as for example in some arcid, cardiids, venerids, and lucinids. Calcium-inefficient designs in bivalves include flat shells and long narrow shells. These less compact forms should appear mainly in environments where the costs of shell-building are lower, whereas spherical shapes should be especially characteristic of molluscs in cold, deep, and fresh waters and of those terrestrial environments where calcium is hard to come by.

These predictions are only partially confirmed by an analysis of shell shapes. In a study of gastropod shell shape in relation to latitude, Richard Graus found that most shells from cold waters conform to the shapes expected on the basis of relatively high building costs. Their apertures are generally round to oval, and unusually high-spired shells are poorly represented among cold-water snails. There is a much greater geometric diversity among tropical and warm-temperate marine snails, an observation that would be consistent with the idea that constraints and calcification are relaxed in warm water. There are many exceptions to this pattern, however. Resorption of inner shell walls would enable snails to recycle precious calcium carbonate, yet groups with extensive powers of resorption—neritids, conids, ellobiids, and many muricids, to name a few—are poorly or not at all represented in cold waters. Turreted shells may be unusual in cold seas, but they do occur. Examples include epitoniid wentletraps, many eulimids (parasites on echinoderms), and the northern-hemisphere turritellid *Tachyrhynchus*. Narrow-apertured shells are also poorly represented, but a few eratoids, triviids, and marginellids with cowrielike shells and slitlike adult apertures occur in cold northern waters, and narrow-apertured volutomitrids are characteristic of polar and deep waters. Lakes and streams, where calcium concentrations are sometimes low, are home to many high-spired thiarids and pleurocerids, as well as to some open-coiled snails such as *Baicalia*, endemic to the cold waters of Lake Baikal in central Asia. Other open-coiled gastropods, whose calcium efficiency is very low indeed, occur in the cold deep sea (some epitoniids, vanikorids, and peltospirids, for example) and on land. Although these evolute shells are thin and delicate, there is no evidence that they are thinner than the more compact shells occurring with them.

No comparable survey of bivalved shells has been carried out, but there is no compelling evidence that calcium-efficient shells are characteristic of cold waters or that calcium-inefficient shells are confined to warm seas. Almost spherical lucinids, thyasirids, cardiids, and venerids occur worldwide in shallow as well as deep waters. Calcium-inefficient flat-shelled isognomonids, pteriids, pectinids, and tellinids are well represented in the tropics, but many cold-water species of tellinids and pectinids also have strongly compressed shells. If the energetics of calcification has set limits on the range of shell forms, these limits are neither absolute nor narrow.

Solutions to Dissolution

Calcium carbonate is a good material for an external skeleton because it is relatively resistant to corrosion. In cold water as well as under high pressure, however, shells built of calcium carbonate are susceptible to dissolution. Although this problem cannot be eliminated entirely, molluscs and other shell-bearing animals have evolutionarily explored several partial solutions, which have important ramifications for other life functions.

One such solution is to make the shell of calcite. In nature, calcium carbonate comes in two principal mineral forms, calcite and aragonite. Compared to aragonite, calcite is slightly harder, less dense (2720 kg/m^3 instead of 2930 kg/m^3), and 36% less soluble.

Several cold-water groups do indeed have calcitic shells. Among periwinkles of the family Littorinidae, for example, the cold-water northern-hemisphere genus *Littorina* and the southern-hemisphere genera *Pellilitorina* and *Risellopsis* have calcitic shells, in contrast to the aragonitic shells of such warm-water groups as *Nodilittorina*, *Tectarius*, *Littoraria*, *Bembycium*, and *Peasiella*. Dog whelks of the muricid genus *Nucella* live in the cool-temperate northern hemisphere and also have shells of calcite. Heinz Lowenstam's classic work on mussels (*Mytilus edulis*) reveals that the outer calcite layer comprises a higher proportion of shells in cold-water individuals than in warm-water specimens.

It would be wrong to conclude, however, that calcite is typical of cold-water shells. Not only are many gastropods and pelecypods in cold waters wholly made of aragonite, but many oysters, scallops, and limpets from warm regions have calcitic shells.

A second way of preventing shells from dissolving or corroding in cold water is to shield the mineral component with an impervious external organic layer or periostracum. Many who have worked with cold-water shells have marveled at the great thickness of the periostracum, which may be velvety, leathery, or even shaggy. The Antarctic trichotropid gastropod *Trichoconcha*, for example, is like a ball of wool to the touch, with so little mineral in the shell that the latter is a flexible husk rather than a rigid shell. Hairy periostraca occur in many families of high-latitude and deep-sea gastropods. Prominent among these are the Trichotropidae, Littorinidae, Buccinidae, Ranellidae, Velutinidae, and the recently discovered Peltospiridae. Many cold-water pelecypods have a thick periostracum that often extends across the commissure. Good examples may be seen in the razor clams (Solenidae), mussels (Mytilidae), Lyonsiidae, Hiatellidae, and Mactridae. The North Pacific chiton *Cryptochiton stelleri* has its large flat butterfly-shaped valves completely encased in a thick rubbery girdle.

The effectiveness of a thick periostracum in protecting the shell from corrosion is dramatically illustrated in the ranellid *Fusitriton oregonensis* (fig. 3.1). Where the periostracum has been worn off, the shells of specimens I collected in Hokkaido, Japan, are so deeply eroded that the apex and several of the early whorls are completely missing. The portions still covered by the periostracum are intact. This example also illustrates the pitfalls of relying on a thick periostracum. If the latter is easily worn off, the underlying shell is subject to rapid deterioration, which could result in higher repair costs and a reduction in overall protection from the outside world.

As with calcite, thick periostraca are not unique to cold-water molluscs, nor do all high-latitude or deep-sea molluscs possess them. Thick velvety or hairy organic coverings are common in tropical ranellids, buccinids, melongenids, conids, and in glycymeridid and arcid bivalves. In contrast to cold-water species with such periostraca, the shells of these warm-water molluscs are typically thick and well calcified. If an organic layer works so well in protecting the mineral component beneath, it is surprising that cold-water molluscs have not extensively gone into partnership with animals that encrust shells. Sea anemones, hydroids, sponges, tunicates (sea squirts), and bryozoans (moss animals) could in principle shield shells from the corrosive effects of cold water, yet few cold-water molluscs are thus encrusted. Some calcitic scallops in the Puget Sound region of Washington and in Europe are covered with sponges, but this cover is effective mainly against predatory sea stars. Indeed, scallops not covered by sponge show no sign of unusual dissolution. Intimate associations between encrusters and hermit crabs are well known and are probably effective in slowing the deterioration of shells. A potential advantage of encrusters over a periostracum is that, whereas the latter cannot be replaced once it is worn away or damaged, encrusters can regenerate and therefore maintain the protective shield. This advantage is especially important to hermit crabs, which cannot repair their shells.

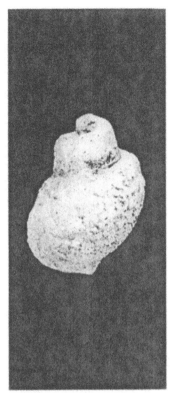

Fig. 3.1. Left, *Fusitriton oregonensis*, Amchitka Island, Alaska. The hairy periostracum typical of this species is largely intact. Right, *F. oregonensis*, Chinbe No
Hana, Hokkaido, Japan. The periostracum has been worn away, with the dramatic consequence that bioerosion has severely damaged the shell. Much of
the spire and part of the anterior end are removed. Remarkably, the animal
was alive at the time the shell was collected. This species attains an average
length of 90 mm.

An alternative to a single external organic coating is a shell wall in
which mineralized layers are separated by organic sheets. This type of
construction is seen in oysters (Ostreidae), basket clams (Corbulidae), and freshwater clams (Unionidae and Margaritiferidae). Such
organic layers probably protect the shells of freshwater clams from
dissolution in the acidic conditions that often prevail in fresh water.
Dissolution of the shell exterior is very conspicuous in freshwater
clams, including some members of the Corbulidae and the related
freshwater family Erodonidae, as well as in freshwater gastropods.
It is curious that calcite occurs in only a few freshwater molluscs

47

(some neritid gastropods, for example) and that, although the organic content of the shell is often very high, hairy or velvety external periostraca are uncommon. Some viviparid gastropods have a periostracum of projecting hairs, but it does not match in thickness the coatings of many marine cold-water shells. Protection of the shell by encrusting organisms is wholly unknown in freshwater molluscs.

Shell dissolution is a potential problem not only in cold or acidic waters. Two types of potential enemy—drilling predators and bioeroders—have the ability to dissolve or to abrade shells. Drilling is a form of predation employed by some gastropods and octopod cephalopods. By applying the rasping tonguelike radula in combination with a softening enzyme, these predators make a small circular hole through the shell wall from the outside. A host of nonpredatory organisms—algae, fungi, foraminifers (mineralized protozoans), sponges, bryozoans (moss animals), polychaete and phoronid worms, barnacles, bivalves, and even gastropods—excavate tunnels or depressions in shells. Although some of these so-called bioeroders attack shells from the inside (the barnacle *Trypetesa* in shells occupied by hermit crabs, for example), most initiate activity on the shell exterior. Both drilling and bioerosion are especially common in temperate and tropical marine waters.

Some of the attributes that protect coldwater shells from dissolution may be effective against biological agents as well. In oysters and corbulid bivalves, for example, the internal organic sheet separating outer and inner mineralized shell layers often prevents naticid and muricid gastropods from completing their drilling. As a result, the frequency of incomplete drill holes in these bivalves is often very high. My former student Miriam Smyth, who studied bioerosion of gastropod shells in Guam, found that stony seaweeds (corallines) encrusting reef-building gastropods effectively prevent the settlement and penetration of borers. If a breach is made in the coralline cover, bioeroders find ready access into the shell. Encrustation by corallines is very common in tropical reef gastropods (figs. 3.2, 3.3). It is even possible that the high relief produced by strong shell sculpture on the exteriors of many of these gastropods promotes settlement of coralline recruits in the low points between the knobs and ridges. All shell-dwelling corallines are, however, also found on adjacent rock surfaces. Any partnership that has developed between snails and corallines is therefore a casual (or facultative) one, and not a specialized or obligate association. Smyth also found that calcitic shells, as well as very hard aragonitic ones, were unfavorable for bioeroders, whereas the softer nacreous structure of trochid snails is more prone to damage from borers.

Fig. 3.2. *Morula granulata*, Pago Bay, Guam. A specimen completely covered with encrusting coralline algae. The encrustation is so thick that the usual sculpture of tubercles is completely obliterated (left, dorsal view). An observer might mistake the shell for a pebble were it not for the fact that the underside reveals a small, neat opening into which a living animal is withdrawn (right, apertural view). This species has an average adult length of 25 mm.

Fig. 3.3. A specimen of *Morula granulata* from Pago Bay Guam, not yet encrusted by coralline algae. Left, dorsal view; right, apertural view.

Structure and Strength

As a protective structure, the molluscan shell must be able to resist breakage. Some of this resistance is imparted by the shell's overall size, shape, and thickness. I shall take up these geometrical aspects of functional design in chapter 7, but here I wish to highlight the contribution that the structure of the shell wall itself makes to strength. Changes that affect the economics of shell production also have mechanical ramifications.

John Currey and his students and colleagues in England have contributed greatly to our knowledge of the mechanical properties of the shell wall. Beginning in the 1970s, they performed a series of tests with pieces of shell from a variety of molluscs representing all the known microstructural types. Resistance to various kinds of stress—tension, compression, and bending—was measured by applying increasing loads in the desired direction on specimens of uniform size and shape.

The results indicated that shell material is anywhere from 3 to 23 times stronger in compression than in tension. In other words, it is easier to break shells by pulling than by pushing. Nacre is generally the strongest form of microstructure in tension, compression, and bending. When nacre does fail under a load, a crack forms and then propagates on an irregular course through the organic matrix rather than through the inorganic mineral tablets. In other types of shell microstructure, cracks travel along smoother paths through as well as between the mineral blocks. Several non-nacreous gastropods and pelecypods have evolved a so-called crossed-lamellar wall structure in which strength of the wall varies according to the direction in which a load is applied. In the gastropod *Conus*, for example, there are three layers of aragonitic lamellae arranged in such a way that shell strength is greater in the spiral direction (perpendicular to the outer lip) than in the direction parallel to the lip. This means that most breaks will remove a sliver of shell at the edge of the aperture, because cracks travel parallel rather than perpendicular to the lip. Damage in the spiral direction is thus relatively minor.

The superiority of nacre in strength applies only to shells that are being constantly maintained by their molluscan builders. Michael LaBarbera at the University of Chicago has found that the nacreous shell of the trochid *Calliostoma ligatum* deteriorates very rapidly after the death of the gastropod. The rapid decay of the matrix, coupled with the high surface area of the remaining mineral component, ensures that the shell loses about half its original strength in compression after only 3 days. The thick shells of other molluscs also weaken after death, but this deterioration is probably slower because of the lower organic content. In order to prevent shells from deteriorating, molluscs must constantly repair small cracks that develop in the shell wall. Such repair may be especially important in nacreous shells, yet it is precisely in these where the costs of repair are high because of their high organic content. Secondary shell-dwellers cannot usually repair their houses. For them, the initial benefits that a nacreous shell might provide very quickly vanish. Because decay is temperature-dependent, deterioration after death may be especially rapid for warm-water shells.

The Shell as a Burden

Movement requires energy. Whenever a shell-bearing animal crawls, swims, leaps, or burrows, the force exerted by muscles or cilia must be sufficient to overcome the inertia in the dead weight of the shell. The heavier the shell, the less rapid will be the acceleration of the moving animal. A major drawback of calcium carbonate and other minerals is weight. Calcium carbonate is almost three times denser than seawater. Tricks to reduce shell weight could therefore enhance an individual's locomotor performance as well as make the shell more suitable for animals that must remain buoyant.

In contrast to calcium carbonate, protein has about the same density as water. An organic-rich skeleton would therefore be lighter, more maneuverable, more suitable for flotation, and altogether less burdensome than a highly mineralized one. Curiously, although many active molluscs have greatly reduced shells or have lost them entirely, those molluscs that retain a fully functional shell do not have have a markedly high organic content in their skeleton. Scallops (Pectinidae), for example, have an organic content of less than 2% in their shells. Most organic-rich shells among gastropods belong to notably slow or even sedentary species. The high costs of producing organic-rich shells may interfere too much with the high costs of locomotion.

Another way to reduce the burden slightly is to build the shell of calcite instead of aragonite. The density of calcite is 7% lower than that of aragonite, a fact that may explain why swimming scallops generally have shells of calcite. However, the shells of swimming limid bivalves, heteropod and pteropod gastropods, and cephalopods are built of aragonite. Even if evolutionarily attainable, the small difference in density achievable by a switch from aragonite to calcite may provide insufficient benefit.

The most readily available and most effective means of reducing skeletal weight may thus not be to modify the shell's chemical composition, but instead to use mechanical tricks for strengthening a thin, light-weight shell composed predominantly of calcium carbonate. I shall discuss these mechanical solutions for reducing the incompatibility between weight and locomotion in chapters 4 and 6.

An Evolutionary Perspective on Cost

Has there been a trend toward cost reduction over the course of the evolutionary history of molluscs? When John Currey and his coworkers first noted the mechanical differences among the various types of shell microstructure, they pointed to the paradox that nacre, the strongest structure, is also a relatively early building material that was generally superceded by mechanically weaker structures in more recently evolved molluscan groups. Palmer's subsequent work on growth rate in relation to the organic content of shells gave further support to this conclusion. Nacre characterizes early gastropod, bivalve, and cephalopod groups. It has been lost in many lineages, especially in those that came to evolutionary prominence during the later Mesozoic and Cenozoic eras. Despite its superiority in strength, nacre is energetically more costly to produce, and shells built of nacre generally grow more slowly and require a longer time for repair. Slow growth may be acceptable and even unavoidable in some environments, such as those in which food is hard to find, but in others it places individuals at potentially high risk. Small individuals are usually more susceptible to predation from a greater variety of would-be enemies than are larger individuals, and small size typically places individuals at a competitive disadvantage. The number of offspring produced by an individual also increases with body size. There is thus a premium on rapid growth in many environments. Moreover, although organic-rich skeletal materials may generally be stronger than are more heavily mineralized ones, they are less hard and therefore more susceptible to abrasion. Nacreous and other organic-rich shells are therefore especially common today in the polar regions, the deep sea, and fresh water, whereas low-cost highly mineralized shells predominate in warm shallow seas.

The molluscan case is not unique. Jennifer Robinson has pointed out that many ancient land plants were rich in lignins, compounds that impart strength to the plant and that have the advantage of being highly resistant to the attacks of herbivores and fungal pests. The disadvantage of lignins, however, is that these substances are energetically costly to produce. Later groups, especially the flowering plants, grow faster and are built with less expensive construction materials and chemical defenses. In the living flora, plants with a low lignin content tend to occur in rich soils and in areas where rainfall and sunlight are plentiful, whereas lignin-rich plants such as conifers and club mosses predominate in poor soils and at high altitudes and latitudes.

Much of our present understanding of cost is based on cool-temperate molluscs. Given the great exuberance of form and sculpture of tropical shells, one is tempted to agree with Richard Graus that the costs of shell construction are lower in the tropics than in cooler marine and fresh waters.

These evolutionary and geographical patterns indicate that reduction in cost may have been important in the long-term economics of life. Two qualifications should, however, somewhat dampen our enthusiasm for embracing this conclusion unreservedly. The first is that estimates of costs in energy and time are still very limited, being available mainly for a few cool-water snails. Absolute costs may vary with habitat and geography in subtle and unsuspected ways. The second qualification is that costs may be acceptable in some circumstances and not in others, according to the magnitude of the potential benefits of greater investment. Costs are not absolute, but relative. Risks and rewards vary geographically, from habitat to habitat, and over time, just as absolute costs of energy and time do. If, for example, there is little risk of greater predation because of slow shell growth, then individuals can remain small for a longer time, so that the costs associated with the production of organic-rich shells may be relatively low even if in absolute terms they are high. Under such circumstances, there may be little evolutionary "incentive" to reduce costs.

It should be clear that an understanding of costs hinges on an understanding of benefits, risks, and rewards. In evolutionary terms, these are translated into success and adaptation. Part II of this book is devoted to these matters.

The following papers and books treat the topics discussed in Part I more fully. The list given here is by no means exhaustive. I have chosen only the most comprehensive, best written, or most insightful accounts. Interested readers can gain entry into the scientific literature by consulting these works. Publications are listed in the same order as the topics presented in chapters 2, 3, and 4.

General Shell Geometry

Ackerly, S. C. 1989. Kinematics of accretionary shell growth, with examples from brachiopods and molluscs. *Paleobiology* 15: 147–164. This paper presents a geometrical framework different from the one offered in chapter 2. It is a good introduction to shell geometry and discusses methods for measuring shell shapes quantitatively.

Raup, D. M. 1966. Geometric analysis of shell coiling: general problems. *Journal of Paleontology* 40: 1178–1190. This classic paper presents four parameters to describe shell shape, and provides a method for comparing observed diversity with what is geometrically possible.

Thompson, D. W. 1942. *On Growth and Form.* Cambridge University Press, London. This is a classic on the mathematics and physics of organic form. In chapter 11, Thompson presents a clear version of gnomonic growth and the logarithmic spiral.

Gastropod Shell Geometry

Gainey, L. F., Jr., and C. R. Stasek. 1984. Orientation and anatomical trends related to detorsion among prosobranch gastropods. *Veliger* 26: 288–298. This paper provides a clear account of how the angle of inclination influences the form of the shell and foot of snails.

Linsley, R. M. 1977. Some "laws" of gastropod shell form. *Paleobiology* 3: 196–206. The author presents several generalizations about shell shape in gastropods and provides a guide to the interpretation of apertural notches, apertural shapes, spire height, and other shell attributes of gastropods.

Vermeij, G. J. 1971. Gastropod evolution and morphological diversity in relation to shell geometry. *Journal of Zoology (London)* 163:

15–23. In this paper, I show how additional parameters of coiling influence the range of geometric possibilities in gastropod shells.

———. 1971b. The geometry of shell sculpture. *Forma et Functio* 4: 319–325. This is a rather abstract account of the geometric properties of ribs, knobs, and spines in shells.

Bivalve Shell Geometry

Carter, R. M. 1967a. On Lison's model of bivalve shell form, and its biological interpretation. *Proceedings of the Malacological Society of London* 37: 265–278. See next citation.

———. 1967b. On the nature and definition of the lunule, escutcheon and corcelet in the Bivalvia. *Proceedings of the Malacological Society of London* 37: 243–263. The two papers by Carter present a coherent overview of shell shape in bivalves.

Rudwick, M.J.S. 1959. The growth and form of brachiopod shells. *Geological Magazine* 66: 1–24. This important paper on bivalve form in brachiopods provides much insight into how growth takes place in shells that are hinged together. It is also the forerunner of the two papers by Carter and of one of my papers (1971b).

Cephalopod Shell Geometry

Raup, D. M. 1967. Geometric analysis of shell coiling: coiling in ammonoids. *Journal of Paleontology* 41: 43–65.

Ward, P. D. 1979. Functional morphology of Cretaceous helically coiled ammonite shells. *Paleobiology* 5: 415–422.

Constraints

Gould, S. J. 1989. A developmental constraint in *Cerion*, with comments on the definition and interpretation of constraint in evolution. *Evolution* 43: 516–539. This paper presents a view of constraint different from the one adopted in this book. Gould believes that constraints imposed by the *Bauplan* are relatively insensitive to ecological change.

Gastropod Torsion and Shell Handedness

Bandel, K. 1982. Morphologie und Bildung der frühontogenetischen Gehäuse bei conchiferen Mollusken. *Facies* 7: 1–198. This monograph presents the most detailed description of torsion in gastropods and discusses in detail the developmental pattern of molluscs.

Freeman, G., and J. W. Lundelius. 1982. The developmental genetics of dextrality and sinistrality in the gastropod *Lymnaea peregra*. *Wilhelm Roux's Archiv* 191: 69–83. This paper describes elegant experiments on the genetic basis of handedness, and summarizes and reinterprets all previous work on the subject.

Gould, S. J., N. D. Young, and B. Kasson. 1985. The consequences of being different: sinistral coiling in *Cerion*. *Evolution* 39: 1364–1379. A careful analysis of shape in left-handed and right-handed snails is presented.

Morris, N. J. 1990. Early radiation of the Mollusca. In P. D. Taylor and G. P. Larwood (eds.), *Major Evolutionary Radiations*, pp. 73–90. Clarendon Press, Oxford. This chapter summarizes our current state of knowledge concerning the early events in molluscan evolution, and presents new interpretations of the origin and significance of torsion.

Pennington, J. T., and F.-S. Chia. 1985. Gastropod torsion: a test of Garstang's hypothesis. *Biological Bulletin* 169: 391–395. This is the only experimental study of torsion in relation to predation on larval gastropods.

Vermeij, G. J. 1975. Evolution and distribution of left-handed and planispiral coiling in snails. *Nature* (London) 254: 419–420.

*Allometry, Resorption, and the Relation between
Growth Rate and Shell Form*

Gould, S. J. 1966. Allometry in Pleistocene land snails from Bermuda: the influence of size upon shape. *Journal of Paleontology* 40: 1131–1141. This paper describes allometry and how to measure it.

———. 1968. Ontogeny and the explanation of form: an allometric analysis. Paleontological Society Memoir 1. *Journal of Paleontology* 42 (part II of 2): 81–98. The hypothesis that change in size with increasing size is adaptive is offered in this paper.

Kemp, P., and M. D. Bertness. 1984. Snail shape and growth rates: evidence of plastic shell allometry in *Littorina littorea*. *Proceedings of the National Academy of Sciences, U.S.A.* 81: 811–813. An elegant experiment is described in which shell shape is linked to growth rate.

Kohn, A. J., E. R. Myers, and V. R. Meenakshi. 1979. Interior remodeling of the shell by a gastropod mollusc. *Proceedings of the National Academy of Sciences, U.S.A.* 76: 3406–3410. This is the most quantitative treatment of internal shell resorption.

McGhee, G. R. 1978. Analysis of the shell torsion phenomenon in the Bivalvia. *Lethaia* 11: 315–329.

Reynolds, P. D. 1992. Mantle-mediated shell decollation increases posterior aperture size in *Dentalium rectius* (Scaphopoda: Dentaliida). *Veliger* 35: 26–35. Not only does this paper present the first evidence of decollation in scaphopods, but it also reviews the phenomenon in gastropods.

Rosenberg, G. D., W. W. Hughes, and R. D. Tkachuck. 1989. Shell form and metabolic gradients in the mantle of *Mytilus edulis*. *Lethaia* 22: 343–344. This short paper summarizes the idea that growth direction is dictated by changes in metabolic rate in the mantle.

Savazzi, E. 1989. Shell torsion and life habit in the Recent mytilid bivalve *Modiolus philippinarum*. *Palaeogeography, Palaeoclimatology, Palaeoecology* 72: 277–282. A good overview is given of twisted bivalves that attach by a byssus.

Signor, P. W., III. 1982. Growth-related surficial resorption of the penultimate whorl in *Terebra dimidiata* (Linnaeus 1758) and other marine prosobranch gastropods. *Veliger* 25: 79–82. This is the most comprehensive paper on internal gastropod shell resorption.

Solem, A. 1983. Lost or kept internal whorls: ordinal differences in land snails. *Journal of Molluscan Studies, Supplement* 12A: 172–178. A useful survey of resorption in land snails.

Vermeij, G. J. 1980. Gastropod growth rate, allometry, and adult size: environmental implications. In D. C. Rhoads and R. A. Lutz (eds.), *Skeletal Growth of Aquatic Organisms: Biological Records of Environmental Change*, pp. 379–394. Plenum, New York.

Evolution of Geometric Diversity

Vermeij, G. J. 1973. Adaptation, versatility, and evolution. *Systematic Zoology* 22: 466–477.

Calcification in Molluscs

Clarke, A. 1983. Life in cold water: the physiological ecology of polar marine ecosystems. *Oceanography and Marine Biology Annual Review* 21: 341–453. This is the most comprehensive account of temperature effects on physiological performance, including calcification.

Graus, R. 1974. Latitudinal trends in the shell characteristics of marine gastropods. *Lethaia* 7: 303–314. This paper presents the hypothesis that shell-shape diversity is related to the energetics of calcification.

Palmer, A. R. 1981. Do carbonate skeletons limit the rate of body growth? *Nature* (London) 292: 150–152. This is an important paper on how the rate of calcification may influence overall growth rate of an animal.

———. 1983. Relative cost of producing skeletal organic matrix versus calcification: evidence from marine gastropods. *Marine Biology* 75: 287–292.

———. 1992. Calcification in marine molluscs: how costly is it? *Proceedings of the National Academy of Sciences, U.S.A.* 89: 1379–1382.

Simkiss, K. 1989. Biomineralisation in the context of geological time. *Transactions of the Royal Society of Edinburgh (Earth Sciences)* 80: 193–199. No other paper discusses the cellular basis of calcification so clearly. It also points out how potentially toxic calcium ions can be removed from cells.

Shell-Wall Microstructure in Relation to Shell Strength

Currey, J. D. 1990. Biomechanics of mineralized skeletons. In J. G. Carter (ed.), *Skeletal Biomineralization: Patterns, Processes and Evolutionary Trends*, Vol. I, pp. 11–25. Van Nostrand Reinhold, New York. This chapter presents an excellent summary of the mechanical properties of shell microstructures and of other forms of mineralization in animal skeletons.

Smyth, M. J. 1989. Bioerosion of gastropod shells: with emphasis on effects of coralline algal cover and shell microstructure. *Coral Reefs* 8: 119–125.

PART II

Life in a Dangerous World:
How Shells Work

The Mechanics of Shells

Shells are built by animals that live in a world of multiple dangers, limitations, and opportunities. Divorced from their natural surroundings, they are objects of abstract architectural beauty in which form takes precedence over function. Only when we observe shells and their makers in nature do we gain some appreciation for the ecological factors that affect the well-being and reproductive success of molluscs. Shells are functional structures whose form reflects the ways in which the animals that build shells are adapted to and limited by their surroundings.

Environments and functions vary greatly from place to place. Temperature, water flow, food, predators, competitors, and a host of other characteristics of an individual's environment vary along gradients of geography and habitat. Shells that work well in one situation may be quite ill-suited to another. The way shells work is, in short, a question of ecology and the adaptive responses of molluscs to it.

I first became conscious of the relation between ecology and form when my family emigrated from the Netherlands to New Jersey in 1955. In the cool wet summers of western Europe, forests and fields were tame places where the worst thing that could happen was to brush against a stinging nettle or a thorny bramble. The grasses and flowers of the polder in which we lived were delicate, gentle plants. The boarding school for the blind I was compelled to attend lay surrounded by quiet woods. Beechnuts and pinecones could be harvested without fear of dangerous animals save for the angry humans looking for a boy who should be doing chores instead. The situation in rural northwestern New Jersey provided a remarkable and exhilarating contrast. Even in the early autumn, when we arrived, days were still warm and the nights resounded with a deafening chorus of crickets. Vines—some thorny, others poisonous—crowded the untidy forest. There were pines with spiny cones, acorns with deep scaly cups, spicebushes and gum trees whose leaves brought forth a spicy odor when rubbed, and coarse hairy weeds that grew tall in the fields. Rattlesnakes and bears were said to be about, and the trees in spring bore huge silken nests of writhing tent caterpillars. Cicadas trilled and oscillated in the trees as the sun blazed in the midday summer's heat.

The forest somehow seemed more alive, more thrilling, and certainly more dangerous than the one I had left behind in the Netherlands.

With the intercontinental contrast fixed in my mind, I was primed to appreciate the differences between the clams from my days at the beaches of the Dutch coast and those beautiful Floridian shells that graced Mrs. Colberg's classroom. Still, without observing the living molluscs firsthand, these differences were hard to understand. Only when I began to visit shores on the Atlantic and Pacific coasts of the United States did the relationships among shell form, function, and habitat first sink in. Looking back, it is surprising that such relatively simple insights should have come only from visits to wildly unfamiliar places. There is, I suppose, nothing like a change in scenery to make understandable the things to which one has become accustomed.

In this chapter I begin an inquiry into how shells work. Molluscs, like other organisms, appear to be adapted to the circumstances in which they live. If shells possess attributes that confer benefits in survival and reproduction to their bearers, it should be possible to identify these attributes, to uncover the principles that connect shell function with the laws of physics, and to document the agencies that cause some individuals to survive and reproduce while preventing others from doing so. In order to understand how the shells of burrowing clams work, for example, we must understand not only how certain shapes and shell textures confer benefits in speed, but also which potential causes of death and reproductive failure would make rapid burrowing beneficial. It is not enough to show that a shell works according to the laws of physics; we must also establish that adherence to these principles confers a biological advantage to individuals.

It turns out to be easier to work out the mechanical principles that underlie biological function than it is to specify which selective agencies bring about adaptation. The principles can be investigated and verified experimentally in the laboratory, whereas the factors that distinguish between the success and failure of individuals must be identified and studied in the wild.

Although these factors are unendingly diverse, they may be usefully classified into two broad categories, which I shall refer to as physical and biological. To the physical category belong those causes of death that are brought about by the unusual and everyday events of weather—storms, fires, earthquakes, floods, excessive heat or cold, lack of water or air, changes in salinity, scouring by ice or sand, and the like. Biological agents—competitors, predators, parasites, and beneficial symbionts—comprise the second category. As will be seen, these agencies cannot always be distinguished easily, nor can it be assumed that the degree of adaptive specialization is determined solely by the physical environment.

In this chapter I consider physical forces and molluscan adaptations to them. Throughout the discussion, I ask how adaptive specialization varies according to geography and habitat. Predators and their techniques are treated in chapter 5; chapter 6 is devoted to the way shells function in defense against predators.

Rigors of the Upper Shore

My first research on shell function was carried out on the upper reaches of tropical seashores. In this no-man's-land, conditions are too salty for all but the hardiest land plants and too hot and dry for most marine life. Rock temperatures as high as 50°C have been measured on the Pacific coast of Panama. When molluscs are exposed to such heat for days without being immersed in water, they may die from thermal stress or from excessive water loss (desiccation). Snails that depend on seaweeds for food have access to only very small plants, which are often plastered as thin films on the rock surface or hidden in self-excavated borings in the rock. For clams that filter plankton out of the seawater for food, the time available for feeding is so short that food intake is limited. Though less extreme, similar circumstances rule the lives of high-shore molluscs on cooler coasts. Despite these rigors, the upper shore is home to a characteristic complement of molluscs, which cope with heat and desiccation in several distinct ways.

Most unusual among these molluscs are the periwinkles or littorines, members of the gastropod family Littorinidae. Nearly every rocky shore, mangrove swamp, and salt marsh supports one or more species of periwinkle. The only shores lacking periwinkles are in the Arctic, Antarctic, and southernmost South America. In the West Indies, where as many as eight littorine species co-occur, the elegantly beaded *Cenchritis muricata* (fig. 4.1) extends so far up the shore that individuals can be found right next to *Cerion* and other land snails.

When the tide is out, most littorines do nothing. The snail's soft parts are retracted in the shell behind a snugly fitting horny operculum, which effectively prevents water loss from the body. The shell is lightly glued by a strand of dried mucus to a rock, branch, or grass blade. In this position, littorines can live in suspended animation for months or even years.

Being unable to move while the foot is withdrawn behind the operculum, littorines are potentially exposed to heating by radiation from

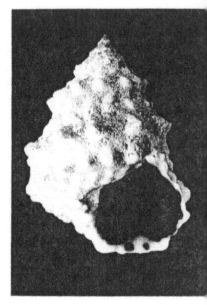

Fig. 4.1. Upper-shore periwinkles from the West Indies. Left, *Cenchritis murica-tus*, near Discovery Bay, Jamaica. The finely beaded sculpture is slightly asymmetrical, with the slopes of the beads being steeper away from the apex than toward the apex. Average adult length is 20 mm. Right, *Tectarius antonii*, Pointe des Châteaux, Guadeloupe. The sculpture consists of sharp knobs set in two spiral rows. This specimen is 18 mm long.

the sun and by conduction from the underlying rock. Attachment by mucus effectively reduces the shell's contact with the rock, and therefore protects the snail from conduction of heat. Only two other options for reducing heat uptake are available to littorines, namely, a decrease in absorption of the sun's rays and an increase in the surface area from which heat can be lost. The shell's shape, color, and sculpture influence the snail's heat budget. In direct sunlight, the amount of heat energy absorbed depends on the surface area of the shell perpendicular to the sun's rays. Heat loss (reradiation), however, depends on total surface area, thus including surfaces that are not at right angles to the incoming rays.

Sun-exposed tropical littorines tend to have light-colored, high-spired, conical shells sculptured by spiral ribs, beads, or small tubercles (fig. 4.1). With the spire pointing in the direction of the sun, only a small portion of the shell surface is perpendicular to incoming radiation. Experiments by Stephen Garrity in Panama show that if a periwinkle is oriented with the shell's spire horizontal, the temperature of the tissues within is higher than in snails oriented with the spire pointing up. A sun-exposed littorine in the latter natural orientation maintains body temperatures 5°–6°C below those of the surrounding rock. The sculpture characteristic of most sun-exposed tropical high-shore littorines creates a large surface area from which heat can be reradiated. Shade-loving species usually have smooth or weakly ribbed shells.

Species occurring lower on the shore are almost always lower-spired than are their higher-shore counterparts, and have smoothly to finely ribbed or weakly granular textures. Away from the tropics, even the upper-shore species are low-spired and weakly sculptured. How are such ecological and geographical patterns to be interpreted? On the one hand, if high spires and sculptured surfaces are adaptive in preventing death from overheating, we might conclude that the importance of heat as an agency of selection increases in an upshore direction as well as toward the tropics. There is indeed evidence that mortality from heat and perhaps desiccation increases upshore. However, why should high-shore littorines at the middle latitudes be lower-spired and less sculptured than their tropical counterparts? Is death from heat and exposure relatively less important to temperate snails? Perhaps there are factors such as strong wave action that select in favor of low-spired smooth shells and that are more important on cooler coasts than in the tropics.

Another intriguing possibility is that the degree of specialization to the rigors of the upper shore depends on which other species are present. The small island of Fernando de Noronha, located about 300 km off the northeastern coast of Brazil, illustrates this point well. Almost uniquely for a tropical island, Fernando de Noronha supports only one littorine species, *Nodilittorina vermeiji*. Instead of being restricted to a single zone on the shore, as most West Indian littorines are, *N. vermeiji* occurs throughout the intertidal. Its highly variable shell is moderately high-spired and has a subdued sculpture of low pointed beads. Despite the fact that Fernando de Noronha is located only 4° south of the equator, its single littorine is an architecturally unspecialized species that combines the features of high-shore and low-shore specialists elsewhere. Given that competition does occur between periwinkle species in other parts of the world, there may be a competitive component to the evolutionary agencies that affect the degree of specialization to physically rigorous environments like the upper seashore.

Although most upper-shore molluscs remain quiescent when the tide is out, few shut down their metabolic machinery to the extent that periwinkles do. Limpets (patellogastropods and lung-bearing Si-phonariidae) and neritids are conspicuous members of the upper-shore community on many shores. While exposed to the air, these snails remain attached to the rock with the foot. Such attachment brings with it the potential disadvantage of allowing heat to be conducted from the rock to the snail, but it also provides some protection from dislodgement by waves and makes possible a method of cooling that is unavailable to resting littorines. This is evaporation. Cooling by the loss of water from the foot and from the space between the foot and the shell works as long as there is a sufficient water reservoir. In the cap-shaped shells of limpets, the size of the reservoir is proportional to the lateral area of the shell, that is, the surface area of the cone apart from the base. In nerites, the reservoir is much larger, because the shell is vaulted instead of conical (fig. 2.14). In high-shore species, in fact, the shell is nearly spherical in shape. Moreover, the inner walls that in most other snails make the shell spirally coiled within are largely resorbed in nerites, so that the shell interior is a single large chamber partially divided by a wall (fig. 2.14).

As in littorines, there are distinct geographical patterns of shell form in limpets and nerites that may reflect the increasing importance of heat and desiccation toward the tropics and in an upshore direction. Nerites living in the uppermost zones of tropical shores are almost spherical in shape (figs. 4.2, 4.3). The shell is typically strongly ribbed and the aperture is small. Lower-shore species tend to have flatter shells with larger apertures and variably developed spiral sculpture. Unlike limpets and littorines, nerites are essentially tropical and subtropical in distribution. A few species penetrate to the warm-temperate shores of South Africa (*Nerita albicilla*), Japan (*N. japonica*) and Australia and New Zealand (*N. atramentosa*). These have a broad distribution on the shore and lack the sculptural and shape specializations of the high-shore tropical species. Some tropical species also have only one intertidal nerite on the rocky shore. Such species—*N. ascensionis deturpensis* at Fernando de Noronha and *N. senegalensis* in West Africa, for example—are broadly zoned species combining the shell features of low-shore and high-shore specialists on other shores. This situation thus closely parallels that seen in littorines.

Limpets occur on almost all rocky shores. Upper-shore species are in the form of tall cones, in accordance with the expectation that they should possess a relatively large reservoir of water and a shape that radiates most incoming heat from the shell. Many lower-shore limpets have very flat shells, but some striking exceptions occur. The north-

Fig. 4.2. *Nerita (Ritena) plicata*, Sella Bay, Guam. This is a typical high-shore species, with a globose shell, small aperture, strong spiral cords, and well-developed teeth on the outer and inner lips. Left, apertural view; right, dorsal view. This specimen is 30 mm in diameter.

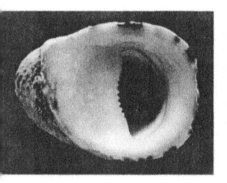

Fig. 4.3. *Nerita (Theliostyla) albicilla*, Miyake-Jima, Japan. This middle to lower intertidal species has a relatively flattened shell and relatively broad aperture. Left, apertural view; right, dorsal view. This species can reach a size of 32 mm in diameter.

eastern Pacific *Acmaea mitra,* in which the height of the shell is about 70% of the mean diameter of the nearly circular base, is a low-shore species often found in the abandoned excavations of sea urchins.

Among high-shore limpets, there is a tendency for the tallest cones to occur on tropical shores. The ratio of height to mean diameter in tropical high-shore specialists is about 0.53, whereas that of similarly zoned warm-temperate species is 0.45. However, this difference is to some extent compensated for by an unexplained difference in shell shape. High-conic tropical limpets have the apex situated directly above the geometrical center of the base (fig. 2.6), so that the shell closely approximates a right cone, whereas most temperate limpets living in the upper intertidal zone on temperate shores have the apex very far forward in position. This difference in form occurs regardless of which major group of limpets happens to be represented. For example, *Helcion pectunculus* in South Africa (a patellid), *Lottia digitalis* in the northeastern Pacific (a lottiid) and *Scurria orbignyi* in temperate western South America (another lottiid belonging to a different subfamily) all have a strongly eccentric apex. It is possible that the right-conical form of tropical high-shore limpets is effective in reducing heat uptake and in increasing reradiation. The light color and strong ribbing of many of these tropical species reinforce such an interpretation, but in our current state of ignorance such ideas are little more than educated guesses.

Life in Waves and Currents

As one descends from the upper beach to parts of the rocky shore that are covered by water at least once a day, the dry and hot conditions are replaced by rigors of a different kind, associated with waves and powerful currents. Molluscs living in such environments face the problem of being swept away. When water motion is severe, molluscs can do little but to shelter in depressions or to cling motionless to stable surfaces. For snails and chitons that must move in order to feed, waves and currents pose a risk because attachment is less secure for an animal in motion than for a stationary one. Accordingly, opportunities for feeding may be few and far between for wave-swept animals even though the surf-beaten rocky shore is typically a food-rich productive environment.

The form and behavior of molluscs exposed to strong flow are dictated in part by the principles governing the flow of water across rigid surfaces. Mark Denny at Stanford University's Hopkins Marine Station has made particularly impressive progress in measuring the forces to which shore organisms are exposed and in identifying ways in which these forces can be counteracted or reduced.

Organisms living in the surf zone are subject to water flow of very high velocity and acceleration. Denny estimates that a wave 2 m high carries water at a velocity of 8 m/sec and an acceleration of 400 m/sec/sec. Under such conditions, organisms must contend with very strong forces. The quantity called force is defined as the product of mass and acceleration. The standard unit of force is the newton, where 1 newton (1 N) is defined as carrying 1 kg with an acceleration of 1 m/sec/sec.

Waves and currents impose three strong forces on organisms. Two of these, drag and acceleration reaction force, act in a direction parallel to flow. For a mollusc clinging to a horizontal surface, these forces create a horizontal stress on the adhesive by which the organism is attached. The third force is lift, which acts perpendicular to flow. It tends to pull upward on a horizontally attached shell.

Two kinds of drag are important. Friction drag is the resistance to flow as water passes over the surface of an organism. It is the rush of water over one's skin during swimming. Pressure drag arises from the reduction in pressure between the upstream and downstream ends of the attached organism. It is the force created by the wake as water fills the space behind an organism in flow. The magnitude of friction drag depends on surface area, surface texture, and stickiness (viscosity) of the water. When a viscous fluid like water flows over an object such as a shell, the layer of water right at the surface remains essentially stationary. The higher the viscosity, the thicker is this stationary layer. Water flowing over a shell can be thought of as a stream of particles, each of which follows a particular path. On the upstream side, the particles follow the shell's contours, but eventually they separate from the shell's surface, so that water from the side or from the downstream end must fill the void. This creates the wake, the size of which determines the magnitude of pressure drag. If the particles trace smooth regular paths as they pass over the object, flow is said to be laminar. By roughening the object's surface or by increasing the velocity of flow beyond a certain critical threshold, this orderly pattern is suddenly replaced by turbulent flow, in which particles of fluid take erratic paths. Under conditions of turbulence, friction drag is increased, but pressure drag is often reduced. Streamlined objects, in which the upstream end is rounded and the downstream end tapers to a thin edge, enable particles of fluid to trace smooth paths that do not separate from the object. In such objects, therefore, both friction drag and pressure drag are low. Lift pulls organisms up from the surface to which they are attached. It is created whenever the pressure of fluid on the upper surface of the object is less than the pressure on the lower surface. Pressure is measured as force per square meter. All attached objects in flow are subject to lift, because the parts of the organism farthest from the point of attachment are exposed to the highest velocity and therefore the lowest pressure. Lift depends on the area of the object projected parallel to the direction of flow. If this area is small, as it is in an upright pole, for example, lift will be low, but drag will be high because the pole has a large cross-sectional area perpendicular to flow. Lift and drag together can cause objects to topple over.

When an organism is hit by a wave, the usual forces of drag and lift that act under conditions of constant velocity are joined by a third force. The acceleration-reaction force is imparted only under conditions of acceleration, that is, when there is a change in velocity. A shell in the path of a wave takes up volume that would otherwise be occupied by accelerating water. The space immediately next to the organism, where flow would normally be reduced, would also be taken up with accelerating water if the shell were not in the way. The mass of water whose acceleration is being impeded by the shell is therefore the sum of the mass of the water directly displaced by the shell's volume and the so-called added mass of water whose flow is influenced by the presence of the shell. Obstacles of high drag usually also have high relative added mass and are therefore exposed to high acceleration-reaction forces.

All these forces must be resisted by structures with which the organism is attached to the shore. In chitons, limpets, and coiled snails, mucus produced by the foot provides adhesion. In mussels (Mytilidae) and some other bivalves, attachment is effected by a flexible byssus of protein fibers produced by a gland in the foot. Other bivalves, notably oysters and chamids, attach one valve to the rock with a cement of calcium carbonate. Cementation is also the means of attachment in vermetid snails. Their irregularly coiled shells are wormlike tubes that often form large aggregations on wave-swept rock platforms.

Denny has estimated the forces to which the Californian limpet *Lottia pelta* is exposed, as well as the resistance this limpet can exert against these forces. The maximum force imposed by waves on *L. pelta* is about 3 N. The limpet is neither especially well streamlined nor especially susceptible to lift. The tenacity of the foot when subjected to an upward-acting pull is between 6.8×10^4 and 2.05×10^5 N/m^2. This means that for a foot whose attachment area is 4×10^{-4} m^2 (or 4 cm^2), the resistible force (tenacity times foot area) is 26–82 N. In other words, the limpet's resistance is very much greater than the maximum force that waves are likely to exert on a stationary limpet. For an actively crawling limpet, tenacity is substantially less (5 x 10^4 N/m^2) than for a stationary one. Strong wave action therefore interferes with a limpet's movement, and effectively restricts opportunities for feeding and for other activities that require the limpet to move from place to place on the rock.

The geometric distribution of shells of wave-exposed intertidal molluscs reflects the ways in which molluscs reduce their exposure to the forces of flow or increase resistance to these forces. These same forces also exclude several architectural types as inappropriate in habitats characterized by strong waves and currents.

Small size is an excellent all-purpose solution to coping with strong forces in flow. All three component forces—drag, lift, and acceleration reaction—have a relatively greater impact on large than on small objects. Moreover, because no part of a small organism extends very far above the attachment surface, water movements and the forces they create are weaker than they would be higher off the bottom. This may explain why tiny snails like *Nodilittorina meleagris*, whose smooth low-spired shells are less than 5 mm in length, thrive on wave-exposed rocks in the West Indies and West Africa. The aperture and foot of this species are not notably large, but the very small size of the snail means that the forces to which the snail is exposed simply are not terribly strong.

Flattening of the body is another obvious means by which the advantage of remaining in the water layer of reduced velocity can be achieved by an organism attached to a wave-exposed surface. For limpets and chitons of this shape, moreover, the area of attachment by the foot is large, and the cross-sectional area projected perpendicular to the direction of flow will be low. As a result, the shell is exposed to relatively low pressure drag. However, the same large basal area has the disadvantage of creating a strong lift force, for this area is parallel to the direction of flow. For the large mid-intertidal Californian limpet *Lottia gigantea*, whose relatively flat shell has a ratio of height to diameter of 0.2 to 0.3, Denny calculates that lift force exceeds pressure drag by a factor of 2.5 or more when flow is from the anterior edge, near which the apex is situated, to the posterior end of the animal. Many low-shore limpets such as the Indo-Pacific *Siphonaria atra* and *Scutellastrea flexuosa* and the West African *Patella safiana* are strongly flattened, having a ratio of height to diameter of 0.3 or less. Some large species, however, have shells in the form of tall cones whose height may be half or more of the diameter of the base. Examples include *P. barbara* in South Africa and *P. laticostata* in temperate Australia. Shells of this shape have high pressure drag but relatively low lift.

Denny's calculations and force measurements show that *L. gigantea* is typical of molluscs attached to wave-swept surfaces in that friction drag and acceleration reaction are much less important than is pressure drag, which in turn is subordinate in importance to lift. This is one reason why few intertidal molluscs and few molluscs living in fast-flowing streams have streamlined shapes. Another reason is that flow varies greatly in direction as well as in magnitude. If flow were consistently in one direction, an organism could orientate in such a way as to minimize drag. This is difficult for animals like limpets that move from place to place in habitats where wave motion can come from almost any direction. Limpets living and feeding on the blades of seaweeds or the leaves of sea grasses are among the few conspicuously streamlined intertidal molluscs. Because the plants on which they live are flexible and therefore move as water flows over them, limpets are consistently oriented in flow. Species such as *Helcion pellucidus* on European kelps and *L. insessa* on the Californian kelp *Egregia* have smooth shells whose length exceeds width and height by a factor of two or three and whose apex lies well in front of the middle of the shell. These characteristics all indicate a streamlined shape.

Gravitational stability is an aspect of shell form that has been little investigated in intertidal molluscs. Drag and lift together can cause organisms to come loose and then to topple. Many gastropods and chitons, however, have a shape that confers a reasonable stability to the animal if the latter is dislodged. In order for an animal to remain in a position with the aperture facing down, its center of gravity must be low and must lie over the aperture. Moreover, the fulcrum (or point on the edge of the base around which the shell will rotate as it falls over) must lie far from the point at which force impinges on the shell. In other words, the lever arm or moment arm of the dislodging force must be long.

If a snail shell is placed aperture-down on a horizontal surface and is then gradually tilted in one direction, it will fall over when a critical angle of tilt is reached. This angle is a rough measure of the shell's stability. In a series of trials with West Indian snail shells, I measured stability in two directions. Posterior instability occurs when the shell tips over backward toward its apex, whereas left-lateral instability occurs when the shell rolls over in a direction away from the outer lip. Wave-exposed species require a minimum of 20° tilt to tip over backward and a minimum of 12° to roll over laterally, but stability is usually much higher. Low spires, especially when coupled with a large aperture, confer high stability. Figures 4.4–4.7 show gastropods characteristic of wave-swept shores. Many muricids, ranellids, and bursids achieve lateral stability by the presence of a varix on the ventral surface of the shell opposite the outer lip (plates 1, 2). Shells of species living on sheltered shores, under boulders, or on sand and mud need not be gravitationally stable and therefore show a much wider range of shapes. Species of *Cerithium*, for example, drag their shells over the rock or sand on which they crawl.

Some snails whose shells have low stability nevertheless live on wave-exposed shores by sheltering in crevices and depressions. This is particularly common in the tropical Pacific and Indian oceans, where many narrow-apertured species of *Conus* (Conidae), *Strigatella* (Mitridae), *Pusia* (Costellariidae), and *Engina* (Buccinidae) are abundant on wave-swept reef flats.

Some chitons and limpets are protected from dislodgment by occupying self-made depressions or home scars in the rock surface. Examples occur in the chiton genera *Acanthopleura* and *Nuttallina* and the limpet genera *Patelloida*, *Scutellastrea*, and *Macclintockia*.

Remarkably, all species with this habit are tropical or warm-temperate in distribution. If dislodgment by waves were primarily responsible for the evolution of the ability to excavate depressions, we would expect to see many cool-temperate chitons and limpets with this ability, especially because the frequency and intensity of storms and powerful waves are greater at higher latitudes. The fact that only warm-water species occupy self-made home scars indicates that factors in addition to waves are at work. Anyone who has ever tried to collect chitons and limpets from home scars will appreciate how difficult it is to dislodge the animals. Observations by David Lindberg at the University of California at Berkeley and by others reveal that predators are more important agents of dislodgement than are waves. These predators include muricid snails, sea stars, crabs, oystercatchers, and fishes. Moreover, chitons and limpets are susceptible to being dislodged by herbivorous fishes and sea urchins that scrape rock surfaces. With the exception of sea stars, these enemies are more numerous and more powerful on warm-temperate and tropical shores than in colder seas.

A similar situation occurs in cemented molluscs. Cementation is an effective means of attachment in waves and currents, but it also works well against scraping herbivores and against many kinds of predators. Some cemented bivalves extend into cold waters (the oyster *Ostreola conchaphila* and scallop *Crassidoma gigantea* in the northeast Pacific,

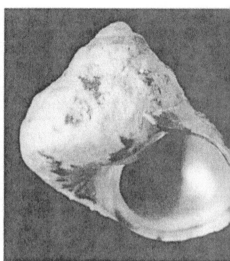

Fig. 4.4. (left) *Agathistoma excavata*, Basse Terre, Guadeloupe. The shell's center of gravity is low by virtue of the large umbilicus, large aperture, and low spire. As a member of the family Trochidae, this species has the axis of coiling steeply inclined to the plane of the aperture. In life, therefore, the apex points up when the snail clings to a horizontal surface. This specimen is 19 mm in diameter.

Fig. 4.5. (right) *Cittarium pica*, Pointe des Châteaux, Guadeloupe. This trochid also has a large aperture, but the umbilicus is relatively small, and the axis of coiling is less steeply inclined to the plane of the aperture. This particular specimen is 24 mm in diameter, but this species can exceed 10 cm in diameter.

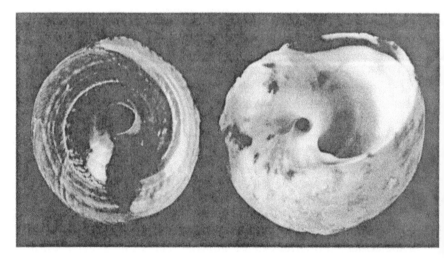

Fig. 4.6. Basal view, *Agathistoma excavata* (left) and *Cittarium pica* (right), showing the large umbilicus of the former and the smaller one of the latter.

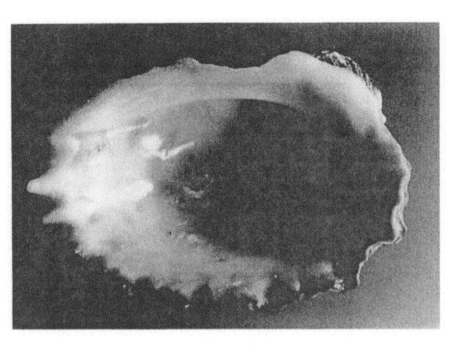

Fig. 4.7. *Concholepas concholepas*, Montemar, Chile. The limpet form is common among snails on wave-swept shores. This remarkable muricid is effectively a limpet. Top, apertural view; bottom, dorsal view. The species reaches a length of 110 mm.

for example), but the great majority of bivalves that attach by cementation live in warm water. More than one-third of the bivalve species found on hard surfaces in the tropical western Pacific are cemented. Worm snails (Vermetidae) are also an overwhelmingly warm-water group.

We see from these observations that the expression of traits reducing exposure to the strong forces of flow or increasing resistance to these forces is affected by circumstances in addition to the flow regime. Just as on the upper shore, the degree of specialization to the physical rigors of the lower shore is dictated by biological factors such as predation, grazing, and competition.

Life as a Swimmer or Floater

Animals that swim or float are exposed to the same forces with which attached shore animals must cope, but their problems are very different. The chief obstacles to a life suspended in water are sinking and instability. At a density of 1026 kg/m^3, seawater is as dense as nonmineralized organic tissues, but less than half the density of most skeletal materials. Most suspended animals therefore have a tendency to sink to the bottom. The rate of sinking scales with surface area, that is, with the two-thirds power of mass. At very small sizes, animals can remain suspended because of the presence of small eddies that counteract gravity. Spines, shields, and other projections can slow the rate of sinking, but these devices are not effective at larger sizes, and no organism more than 2×10^{-9} kg in mass can rely on eddies to remain suspended. This corresponds to a diameter of about 150 microns (0.15 mm), well within the range of sizes of swimming larvae but smaller than most adults.

Two mechanisms prevent sinking of large animals. One is to reduce the animal's density by eliminating heavy skeletal material or by incorporating a low-density substance such as a gas. The animal's overall density (mass per unit of volume) will then be the same as that of the water in which it swims. The second mechanism is to create upward-acting thrust and lift by the use of organs of propulsion such as fins, tails, and devices that expel water in a jet.

Swimming animals have more than sinking to contend with. They are subject to the same forces in flow—drag, acceleration reaction, and lift—that stationary organisms are in waves and currents. Friction drag and pressure drag act in a direction opposite to motion, and can be reduced by streamlining the body. Acceleration reaction is the resistance to changes in an individual's velocity. When the animal is speeding up, acceleration reaction acts in a direction opposite to motion; when it slows down, acceleration reaction acts in the direction of movement. It also tends to prevent sudden changes in direction and thus interferes with maneuverability. Ways to reduce acceleration reaction include streamlining, a reduction in the mass of structures not producing thrust, and a reduction in the mass of water carried along with the moving animal. Maneuverability is enhanced in a short, laterally compressed body. The thrust provided by muscles in the organs of propulsion must overcome the resistances imposed by drag and acceleration reaction. Upward-acting lift in fixed organisms can cause dislodgement and is therefore a threat, but for swimmers it prevents sinking and should therefore be enhanced. Structures creating lift in water are known as hydrofoils. Lift results whenever the pressure below is greater than the pressure above. This happens when the velocity of flow across the lower surfaces is less than that on the upper surface. The difference in pressure that creates lift also creates some drag. To reduce this so-called induced drag, hydrofoils must be broad in the direction perpendicular to motion and short in the direction parallel to motion.

Passive floating is a relatively uncommon mode of life in snails. Purple sea snails of the family Janthinidae float in the open-ocean plankton, where they feed on jellyfishes. Many small snails have the ability to hang at the water surface from a mucus thread, and can sometimes take advantage of the surface tension to remain afloat. Freshwater pulmonates often bob up from the bottom of pools and ponds to take in oxygen at the water surface. All these snails have very thin, smooth shells, but they lack specializations for buoyancy.

Cephalopods with an external shell rely on a low-density gas in the chambers of their shells to avoid sinking. This gas is maintained at a constant pressure of 900 kg/m^2, which is slightly less than air pressure at sea level, irrespective of whether the animal is at the water surface or at greater depths. Because pressure on the outside increases with depth, the shell will implode below a critical depth. The volume of gas in the shell is controlled by the siphuncle, a tubular organ that extends through perforations in the septa to all the chambers of the shell (fig. 4.8). When each chamber is formed by closing off the hind part of the body chamber with a septum, the newly set-off chamber is full of liquid. This liquid is gradually pumped out by the siphuncle and replaced with gas. When the animal descends, the outside pressure rises while that of the gas inside the chambers remains the same. The difference in pressure between the outside and the inside of the shell must be resisted by the shell wall, the septum, and the siphuncle.

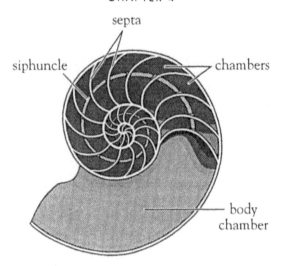

Fig. 4.8. Cephalopod shell showing, siphuncle, chambers, septa, and body chamber.

Resistance to a difference in pressure is achieved by strengthening the shell and the wall of the siphuncle. One way to increase a shell's strength is to thicken the shell wall. This, however, not only makes the shell denser, requiring a greater volume of gas in compensation, but also contributes to higher acceleration reaction force during swimming. Acceleration reaction is important in cephalopods, which swim by forcing a jet of water out of the funnel, an extension of the mantle. Each time water is ejected, the animal speeds up slightly, and this acceleration is resisted by the mass of the shell. Accordingly, a lightweight, thin-walled shell that is strengthened by corrugation of the outside wall and by internal supports (the septa) is better suited to fulfill the requirements associated with swimming and the maintenance of buoyancy than is a thick shell.

A siphuncle of small diameter is ideal for resisting differences in pressure. Unfortunately, such resistance is achieved at the expense of the ability to respond rapidly to sudden changes in pressure brought about by rapid ascent or descent. The thicker the wall of the siphuncle, the slower is the removal of liquid from newly formed chambers. Although this functional incompatibility can be overcome by increasing the diameter of the siphuncle, that solution has the disadvantage of reducing resistance to a large difference in pressure and of taking up space in the chambers that would otherwise be filled with gas. Slow removal of liquid from the chambers is tolerable for a cephalopod that stays within a narrow depth range, but would be unacceptable for an animal like *Nautilus* that undertakes vertical excursions or for an individual whose density changes suddenly. A sudden unanticipated change in density could occur because of damage. A puncture in the shell, for example, could flood one or more of the chambers, resulting in a sudden increase in weight; the loss of a piece of shell at the outer lip, on the other hand, would reduce density and cause the individual to rise to the surface. Rapid response to such changes in density is therefore important. Such compensation is possible only when the siphuncle is thin-walled or large.

The demands of buoyancy control and of swimming combine to favor thin but strong shells with a thin-walled siphuncle. Streamlining and high maneuverability are made possible by a laterally compressed shell. In chapter 6, I consider how cephalopod shells can be both strong and thin.

Scallops (bivalves of the family Pectinidae) also swim by jet propulsion, but they lack a low-density fluid or gas to achieve buoyancy. For this reason, scallops do not spend all their time swimming, but instead often lie on the sea bottom.

The scallop's shell takes an active part in swimming (fig. 4.9; plates 3–5). Thrust is created by the contraction of the large centrally located adductor muscle. One end of the muscle is inserted on the center of the inner surface of one valve, and the other end is inserted slightly more ventrally on the other valve. As the adductor contracts, the valves are clapped together, and water is forced out between the valves at four points: in front of the hinge line, behind the hinge line, and at the front and back ends of the main disc of the shell. In a swimming scallop, the direction of locomotion is toward the ventral edge of the valve and away from the hinge, which is therefore situated at the back of the moving animal. When the scallop is at rest, it lies on its right valve.

Some relatively unspecialized scallops can swim obliquely upward for short distances and then resume their normal life on the sea bottom. One or both valves tend to be strongly inflated and to possess well-developed radial folds that impart rigidity to the otherwise thin valves. On the bottom, the animals are attached by a byssus extruded through a notch in front of the hinge line in the right valve, or they lie free in sand and mud (fig. 4.9; plates 3–5).

The ability to swim longer distances has evolved several times among scallops. Itaru Hayami at the University of Tokyo has intensively studied swimming in scallops of the genus *Amussium* (plates 3–5). Speeds of up to 1.6 m/sec have been observed in the Western

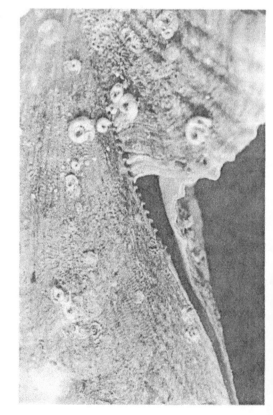

Fig. 4.9. *Swiftopecten swiftii* (*Pectinidae*), Akkeshi Bay, Hokkaido, Japan. The valves of this relatively sedentary species are strongly convex, and bear well-developed radial folds. The scallop spends much of its time attached to rocks by a byssus, which is located in a notch in front of the anterior ear or auricle on the right valve. Top, shell exterior; bottom, close-up of the byssal notch. The actual specimen is 73 mm.

Australian *A. balloti*, which can swim for distances of 23 m or more. The thin platelike valves of *Amussium* are strengthened on their inner surfaces by widely separated radial ribs, but externally they are smooth and shiny, so that friction drag is low. At the points where water is ejected as the valves are shut, there are permanent gapes between the valves. Although the leading ventral margin of the shell is sharper than it would be in an ideally streamlined hydrofoil, maximum convexity occurs far toward the leading edge in accordance with the requirements of a streamlined object.

Gastropod swimmers use extensions of the foot as hydrofoils and therefore depend on lift, sometimes in combination with gas in a float, to stay off the bottom. They include predatory heteropods and pteropods, which spend all their life in the plankton, as well as temporary swimmers that usually live buried in sand. All have thin smooth or weakly sculptured shells.

Life on Sandy and Muddy Bottoms

Most of the world's ocean is floored by sand and mud; so are many lakes, streams, and rivers. Not surprisingly, therefore, sandy and muddy sediments are home to an enormous number of shell-bearing molluscs. Some of these live on the surface of the sediment and are therefore said to be epifaunal. Others live buried in sand or mud and are said to be infaunal.

The challenge faced by epifaunal animals on sand can best be appreciated by standing on a sand beach in the wash of the waves. Normally, the sand is firm, but as a wave passes over one's feet, the sand around them is winnowed away and the feet slowly sink into a deep hole, which is eventually filled in again with sand.

For epifaunal animals on mud, the situation is different. As anyone walking across a mud flat knows all too well, the problem with mud is that the surface yields under weight. Even if the animal were sufficiently buoyant to float on top of undisturbed mud, it may sink if the mud is disturbed by nearby activity.

In order for an epifaunal organism to remain on top of the mud or sand, the stress (weight per unit of area) it applies to the sediment surface must be less than the stress that the sediment surface can bear. Stress on the sediment can be reduced in three ways: by reducing the animal's volume, by reducing the animal's density, and by increasing the area over which weight is distributed. For shell-bearing molluscs, a reduction in density means a decrease in shell thickness, because the shell is the densest part of the body. Scallops living on deep-sea ooze have very thin fragile shells and are generally of small size.

There are two ways to distribute an animal's weight so that the force of gravity at any one point is small. In the snowshoe method, the whole of the animal lies on the surface of the sediment, and the area covered by the animal is large. Among living molluscs, the snowshoe habit is perhaps best exemplified by bivalves such as the Southeast Asian *Placuna placenta* which lies on its nearly flat thin left valve on soft mud. The best example I have come across among gastropods is the freshwater lymnaeid *Radix auricularia* (fig. 4.10) in lakes in Austria. The outer lip of the very thin adult shell is greatly expanded into a winglike horizontal flange, enabling this snail to crawl on the surface of fine mud.

In the second method, the iceberg habit, part of the animal is submerged beneath the sediment surface. The surface area in contact with the sediment is therefore large relative to total surface area of the animal. Only that part of the animal extending above the sediment will bear down on the surface. This mode of life is fairly common among oysters and scallops. Among the latter, the lower right valve tends to be more convex than the upper left valve and is usually partly sunken into the mud or sand. In extreme cases, such as the genus *Pecten*, the upper valve is flat or even concave when seen from above.

Fig. 4.10. *Radix auricularia* (Lymnaeidae), Zellersee, Austria. The lip of the adult snail is expanded into a thin flange that helps support the animal when crawling on mud. Unlike the flared outer lip of many marine snails, that of the freshwater *R. auricularia* evidently keeps growing throughout life. Left, Apertural view; right, dorsal view. This specimen is 29 mm long.

Burrowing

The great majority of sand-dwelling and mud-dwelling molluscs are infaunal. For them, remaining buried in the sediment is a major challenge and a unifying theme of functional design. Because infaunal animals are so conspicuous in the fossil record, these functional aspects have drawn the attention of paleontologists. Steven M. Stanley's landmark studies of bivalves, begun at Yale University, ushered in efforts by Philip Signor at the University of California at Davis and by others to make sense not only of clams, but also of the great diversity of burrowing snails. The results of this work indicate that two basic ways are available to molluscs for remaining buried. These are stabilizing the sediment around the animal and rapid burrowing.

Waves and currents create drag forces that cause grains of sand and mud to be suspended in the water and transported to other sites of deposition. Infaunal animals are thus gradually unearthed, and may themselves be carried off. If they are not attached to some object in the sediment, infaunal animals have only two means available to achieve stability in the face of sediment scour. One is to be buried deeply in the sediment, so deeply that even the strongest currents do not reach the animals or the sediments surrounding them. The northeastern Pacific clam *Panopea abrupta* (the geoduck) buries to a depth of 40 cm or more; the angel wing *Scobinopholas costata* in Florida may be buried at an even greater depth in mud and peat. The second means is to stabilize the sediment next to the shell by means of projecting sculpture.

Among bivalves, deep burial is associated with unsculptured shells and long siphons. Water is brought in from the surface of the sediment by way of the inhalant siphon and expelled after passing through the gills through the exhalant siphon. These tubular extensions of the mantle are typically housed in the posterior part of the shell. Tellinids and some other deep burrowers can retract the siphons fully into the shell when the valves are shut by the adductor muscles. For such clams with retractile siphons, the relative length of the siphons is indicated on the shell by the relative size of the pallial sinus (fig. 2.4). Many other deeply buried clams, however, cannot retract the siphons. This inability is indicated by a permanent gape between the posterior edges of the valves when the shell is closed. Razor clams (Solenidae and Cultellidae), soft-shelled clams (Myidae), and most deep-burrowing Mactridae and Hiatellidae provide good examples of permanently gaping shells (fig. 4.11).

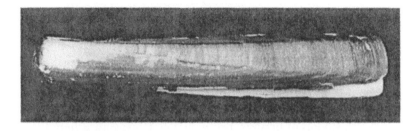

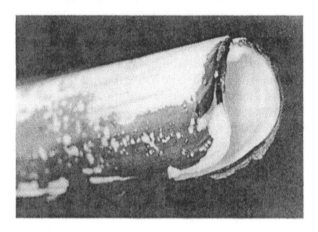

Fig. 4.11. *Ensis siliqua*, Newgale Sands, Wales. This razor clam has a smooth surface appropriate for rapid burrowing; both the front and back ends of the shell gape, even when the valves are shut. Top, External view; bottom, posterior gape. This specimen is 18 cm long.

For shallow burrowers, stability in the sediment must be achieved by modifications of the shell. Several of these involve shape. The hind end of the shell is modified in either of two contrasting ways from the typical tapered shape. In so-called rostrate bivalves, the posterior end is drawn out into a more or less slender narrow projection. It enables the bulk of the body and shell to be located well beneath the sediment surface while only a small part is near the surface. The rostrate condition is seen in many nuculanids, cuspidariids, semelids, and mactrids. The opposite solution is found in truncate bivalves, in which the hind end is laterally expanded and set off from the narrower remainder of the shell by a keel (figs. 4.12–14). Truncation has evolved in numerous freshwater clams as well as in warm-water marine members of the Arcidae, Trigoniidae, Cardiidae, Veneridae, Glossidae, Donacidae, and Mactridae, among other families. Shells with a broadened posterior end occur commonly in tropical assemblages in shallow-water sand (10%–38% of species) and in fresh water (13%–21%), but they are absent at higher latitudes and in the deep sea.

External sculpture also stabilizes shells buried in sediment. Trapping particles between adjacent ridges, a sculpture of ribs reduces movement of sediment around the buried shell. The strong spiral sculpture of turritellid snails, which are more or less sedentary infaunal filter-feeders, probably serves to stabilize the high-spired shells of these snails in sand and mud. Strong concentric ribs are very common in such infaunal clam families as the Nuculidae, Veneridae, Tellinidae, Astartidae, Mactridae, Lucinidae, and Corbulidae (fig. 4.12). Radial ribs are characteristic of arcids, carditids, cardiids, and some venerids (fig. 4.13). Very few freshwater clams display these typical forms of stabilizing sculpture. A few small sphaeriid pill clams have fine concentric riblets, but regularly arranged radial ribs are practically unknown in clams from fresh water. Some Asian and North American genera such as *Lamprotula, Plethobasus, Obliquaria,* and *Cyprogenia* have an unusual ornament on the hind part of the shell consisting of irregularly disposed rows of tubercles. Whether this sculpture serves as a stabilizing mechanism is not known.

Burrowing of molluscs is typically accomplished by the foot. Stanley devised a measure of burrowing rate that is independent of animal size. His burrowing rate index is defined as the cube root of mass divided by the time in seconds required for the animal to achieve complete burial from a starting position at the surface of the sediment. Very slow burrowers are those in which the burrowing rate index is less than 0.1, whereas fast burrowers are those in which the index is 1.0 or higher. Some species of the beach-clam genus *Donax* take only a little more than 1 second to burrow completely, and have burrowing rate indices of 17 or more. The fastest snail burrowers we have seen are members of the Olividae, in which the burrowing rate index exceeds 3.0. South African species of *Bullia* (Nassariidae) may even be faster.

The ability of molluscs to burrow can be enhanced either by decreasing drag and acceleration reaction force or by increasing the size of the organs providing thrust. Drag reduction is associated with streamlining. The outer surface of the shell should be smooth or sculptured in such a way that grains of sediment are channeled in a direction parallel to that of burrowing. This means that sculpture, if present, should be collabral (parallel to the outer lip) in snails and

Fig. 4.12. *Harvella elegans* (Mactridae), Las Lajas, Panama. The posterior end of the shell is set off by a sharp keel and is therefore truncated. Together with the concentric folds or corrugations, which are visible on both the interior (top) and exterior (bottom) of the thin valves, the truncated posterior end stabilizes the animal in the sand. This species reaches a length of 56 mm.

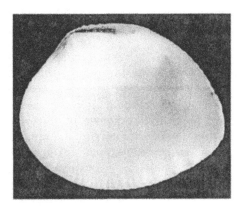

Fig. 4.13. A shallow-burrowing clam, *Gafrarium tumidum*, Babeldaob, Palau. The strong radial ribs may stabilize the shell in the sand in which it is buried. The well-developed hinge and strongly saw-toothed or crenulated margin of the valves make for a tight seal when the valves are shut, and prevent the valves from being sheared apart. Left, Exterior view; right, interior view. This species has an average length of 40 mm.

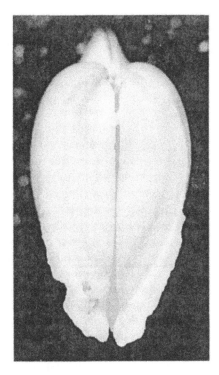

Fig. 4.14. *Mactrellona clisia* (Mactridae), Las Lajas, Panama. This smooth, thin-shelled clam has a distinctly truncated posterior end. Left, Oblique view of the truncated posterior end of the shell; right, truncated posterior end of the shell. This species reaches a length of 73 mm.

usually concentric in clams. The anterior part of the shell, which usually points downward during burrowing, should be wedge-shaped. In the case of gastropods, the sutures (lines of contact between the whorls) should be so shallow that the sides of the spire present a flat or smoothly convex profile without marked break points. Discontinuities would interrupt the smooth flow of particles backward along the shell as the animal burrows, and would cause a large amount of sediment to travel along with the burrowing animal. The same considerations preclude the presence of an umbilicus.

Like all other forms of animal locomotion involving the contraction and relaxation of muscles, burrowing is a cyclical process done in "steps." The foot is first extended downward into the sediment. In clams, this thrust is followed by a quick shutting of the valves, an action that expels water out from between the valves. The shell is then pulled down into the sediment by the contraction of the pedal retractor muscles in bivalves and the columellar muscle in snails. In clams, the front pedal retractor muscle contracts first, followed immediately by contraction of the posterior one. This sequential contraction causes the shell to rock in the sediment. Rocking does not occur in clams with very elongate shells, but it is very conspicuous in species of the genera *Strigilla* and *Divaricella*, whose circular shells may rotate through an angle of as much as 45° within the plane of the commissure. The sequence of foot extension and retraction is repeated until the final burial depth is reached.

The downward thrust of the foot, as well as the shell rocking that results from sequential contraction of a clam's pedal retractor muscles, often causes the shell to slip backward or upward slightly. Many burrowing snails and clams have evolved asymmetrical sculpture that prevents back slippage and therefore reduces the number of "steps" needed to complete burial. In this so-called ratchet sculpture, the side of a rib toward the direction of burrowing has a gentle slope and slides easily into the sediment, whereas the opposite side has a steep slope and resists backward or upward motion. In *Strigilla* and *Divaricella*, ratchet sculpture consists of oblique ribs, those on the front part of the shell being oriented at right angles to the ribs on the posterior part. As the shell rotates in the plane of its commissure during retraction of the foot, the steep dorsal slopes of the ribs are alternately thrust against the sediment. In most other bivalves, the asymmetrical sculpture is in the form of concentric or unidirectionally oblique ribs. In gastropods, all ratchet sculpture is spiral. Some smooth gastropods may achieve the advantages of ratchet sculpture by having asymmetrical sutures, in which the posterior edge of one whorl slightly over-hangs or steps down to the surface of the whorl behind.

Most molluscs with sculpture that enhances the ability to burrow are relatively but not extremely rapid burrowers. The burrowing rate typically varies from 0.3 to 1.0. Only in the tellinid genus *Strigilla* in which the burrowing rate index is about 9, is burrowing genuinely fast. Clams with ratchet sculpture include some cockles (Cardiidae), venerids, tellins (Tellinidae), lucinids, and trigoniids. Among gastropods, it occurs in high-spired cerithiids (*Rhinoclavis*) and some auger shells (Terebridae) as well as in low-spired nassariids, miter shells (Mitridae), helmet shells (Cassidae), and acteonid bubble shells (figs. 4.15, 4.16). Stepped spires are particularly well developed in some Indo-Pacific auger shells, olive shells (Olividae) (plates 6, 7), and miters (Mitridae and Costellariidae).

Ratchet sculpture is particularly frequent in shallow-water sand-dwelling clams. I have observed it in 11%–35% of tropical assemblages of clams, with the highest incidences occurring in the Pacific. Cool-temperate, deep-sea, and freshwater clams rarely or never develop sculpture that enhances burrowing. In gastropods, the phenomenon is even more restricted geographically, being confined almost entirely to the tropical Indo-Pacific region, where up to 16% of shallow-water sand-dwelling species are characterized by ratchet ornamentation.

Although a reduction in drag and acceleration reaction is important in fast burrowers, a large engine—the foot in the case of molluscs—is required to furnish the power. In most bivalves, the foot is extended through the anterior or ventral sector of the shell opening. A large foot is therefore indicated by a large anterior portion of the shell, which in turn means that the umbo is situated relatively far back on the shell. In cockles (Cardiidae), the foot is extended ventrally, and the shell is accordingly exaggerated in a ventral rather than in an anterior direction, that is, the shell of fast-burrowing cockles is higher than it is long. In snails, the size of the foot is usually indicated by the size of the aperture. A large aperture thus indicates a large foot. Exceptions occur in moon snails (Naticidae), olive shells (Olividae) (plates 6, 7), and marginellids, in which the very large foot envelops part or all of the shell. In these snails, the foot can be fully withdrawn into the shell when danger threatens. Because a large aperture is often inconsistent with a shell's function as an armored fortress, rapid burrowing and reliance on armor are generally incompatible except in the groups in which the foot extends over the shell.

Fig. 4.15. *Neocancilla clathrus* (Mitridae), Piti Bay, Guam. The ratchet sculpture prevents back slippage as the snail burrows forward and downward into the sand in which it lives. The narrow aperture protects the snail from attack by predators. Top left, Dorsal view; top right, apertural view; bottom, detail of the ratchet sculpture on the spire. This species reaches 31 mm in length.

Fig. 4.16. *Rhinoclavis sinensis*, Palapala Bay, Pagan, northern Mariana Islands. Although the small aperture and foot preclude the possibility of rapid burrowing in this species, the presence of ratchet sculpture and of almost imperceptible sutures between adjacent whorls makes the shell relatively streamlined. Varices spaced about 225° apart provide strength to the shell and limit the extent to which the outer shell wall can be stripped away by lip-peeling predatory crabs. The recurved anterior siphonal canal enables the burrowed snail to maintain contact with the water above the sand in which the snail is buried. This specimen is 48 mm long.

The Riddle of Specialization

In this chapter we have seen how the laws of physics together with a knowledge of the forces to which organisms are exposed can explain how shells work. By identifying which forces are important and by establishing how each could be counteracted or put to advantage, we can map out pathways of adaptation. This in itself is a rewarding exercise in functional design, and one that is very far from being completed; but if in the end we have learned only that shells work well and do not violate established mechanical principles, we will have done little more than confirm the obvious. After all, shells should be well designed if their makers thrive in habitats where the forces and energy associated with gravity, flow, and exposure to air play important roles.

Of course, the study of functional design in organisms does more. Besides enumerating pathways of adaptation, we can ascertain which factors and circumstances control the "choice" of design. Perhaps more interestingly, we can ask why organisms in some places have proceeded further down a particular path of specialization than have others living in apparently similar physical circumstances. I already made mention of this puzzle in the discussion on adaptation to the rigors of the upper shore. Although heat and desiccation would seem to be important to high-shore snails everywhere, shell attributes enabling snails to cope with them are best expressed in the tropics, especially in the Indo-Pacific region. Dislodgement by waves is an ever-present danger for molluscs on wave-swept rocky shores, yet only on relatively warm coasts do we find limpets that excavate depressions in rock from which it is very difficult to dislodge the occupants. Given that storms may be more frequent as well as more intense at high latitudes, we might have expected to find excavating limpets on cold-temperate shores as well as in warmer waters.

That these examples are not isolated exceptions is indicated by the pattern of distribution of shell traits related to burrowing performance. I have found only one cool-temperate bivalve (the cockle *Cerastoderma edule* in Europe) that occasionally has ratchet sculpture on its shell. No cool-temperate species in the North Pacific or northwestern Atlantic have this sculpture, nor do any have a truncated posterior that enhances stability in the sediment. Ratchet sculpture is also unknown in freshwater molluscs. On tropical shores, 15%–34% of infaunal bivalve species possess ratchet sculpture in assemblages from sand. Smoothly polished clams with a large foot as inferred from a ventrally elongated shell (cockles) or from a shell in which the umbo lies behind the middle do not occur in fresh water. They account for 17% or less in cool-temperate marine infaunal bivalves and for 13%–31% of species in tropical clam assemblages. Yet, waves and currents affect bivalves in rivers, lakes, and shallow-water seas all over the world, as illustrated by the fact that temperate as well as tropical beaches are strewn with shells washed out of the sand just offshore by storms.

Patterns of distribution such as these could be explained in various ways. Perhaps the forces to which shells are adapted are stronger in the marine tropics than elsewhere. Perhaps species in the tropics occupy a narrower range of environments. Thus, whereas temperate species might live throughout the intertidal zone or in the sand as well as on firm surfaces, tropical species would be confined to only one zone or to only one kind of bottom. Specialization of this kind might be enforced by the presence of co-existing competitors or predators. Finally, there may be agencies of selection at work that greatly magnify the nonbiological effects of waves, currents, aerial exposure, and other rigors. Effective burrowing and stability in the sediment, for example, might benefit infaunal molluscs when the cause of exhumation or disturbance is a predator rather than a storm. Anchorage by shell weighting might work well for freshwater clams that are exposed to few burrowing predators, but in the sea there might be a greater premium on escape by burrowing, so that stabilizing devices that interfere less with burrowing performance would be favored over those that rely on a thick heavy shell. The point is that we cannot expect to understand patterns of specialization and choices of adaptive pathways until we learn about the way that other organisms evolutionarily affect molluscs. This topic will be taken up in the next two chapters.

Predators and Their Methods

In the life of all naturalists there are defining moments that shape their conception of how nature works. For me, one of those moments came during an afternoon field trip with my close friend Lucius G. Eldredge, then at the University of Guam Marine Laboratory. Our destination that day in the summer of 1970 was Togcha Bay, a shallow indentation on the windward coast of Guam, in the tropical western Pacific. As the tide was ebbing, more and more of the reef flat was becoming exposed, leaving only occasional pools and sea-grass beds where the water became like a hot bath under the unrelenting sun. As we walked toward the surf-beaten reef edge, Lu paused briefly to pick up a shell. It was a money cowrie (*Cypraea moneta*), still shiny, but with its entire top missing (fig. 5.1). Just one more broken shell, I thought to myself as I casually fingered the abused specimen. "I see a lot of these," I said, more in disgust than with interest. "Yes," Lu replied, "I

Fig. 5.1. *Cypraea caputserpentis,* Pago Bay, Guam. The top of the shell is broken off (left) while the region around the aperture remains intact (right). This kind of damage, which crabs are known to inflict, is very common in cowries. This specimen is 31 mm long; the broken shell wall is 1.1 mm thick.

think crabs do it. I've seen crabs break cowries like this in our saltwater aquarium at home."

Insofar as I had ever thought seriously about shell breakage, I considered it a nuisance, the result of the battering that empty shells receive from being caught between boulders or from collisions with rocks exposed to the surf. To me, broken shells were not worth picking up. The fact that many broken shells are found on sheltered shores, where the force of waves was minimal, had never occurred to me. Neither had the observation that many shells have repaired breaks, meaning that at least some shells sustain damage during the life of the builder. With the realization that predators might be responsible for shell destruction, I began to appreciate that broken shells contain information and should not simply be tossed away as ugly imperfections.

Thus began my research program on the role of enemies in the evolution and distribution of molluscs. First, it was important to identify possible predators. As an assistant professor at the University of Maryland at College Park, I was within easy reach of the U.S. National Museum of Natural History in Washington, home to the largest collection of crustaceans in the world. I soon stumbled onto two crab genera, *Carpilius* and *Eriphia*, that looked like excellent candidates as shell-breakers. Their right claws were huge, and the teeth lining the inner surfaces of the apposing fingers of the claw were massive molar-like thickenings. It took little imagination to suppose that these crabs could be powerful shell-crushers, but imagination was about all I had to go on, for next to nothing was known about the biology of these crabs. Accordingly, I decided to return to Guam, where these crabs were known to occur and where damaged shells were common.

In Guam, my wife Edith Zipser and I began to observe how crabs attack molluscs. We were interested not only in which attributes of shells were effective in thwarting the crabs, but also in the time required by crabs to breach the various shell defenses. Our work soon extended to other predators—fishes, snails, and sea stars, for example—and began to reveal just how important enemies are in controlling the lives of molluscs.

Predation and Selection: The Evolutionary Importance of Failure

At the same time we were listening to crabs break shells in the laboratory and recording which shells withstood attacks and which did not, I began to study the prevalence of breakage in nature. From my previous experience it was already clear that breakage was a common fate of shallow-water marine snails, but casual observations are not enough to prove the point. At specific sites I therefore sampled all shells and shell fragments no longer occupied by the snail builders in order to calculate the proportion of snail deaths that is due to shell breakage.

A high incidence of death by breakage does not, however, necessarily mean that breakage is evolutionarily significant. For example, if all attempts by predators to break shells resulted in the prey's death, the shell would be useless as a means of protection; persistence of the prey population in the face of such predation would occur not by virtue of the protective role of shells, but in spite of it. The only way in which traits conferring resistance to breakage can spread through a population is by the survival and subsequent reproduction of individuals that have successfully withstood onslaughts by shell-breakers. The evolution of antipredatory protection, in other words, is contingent on predator failure.

Military commanders during World War II were aware of this principle when they pondered the question of how to improve the design of warplanes. Many planes were lost during bombing runs over Germany, but others came back damaged by flak. Though damaged, the returning planes had been successful; the pilot stayed alive despite encounters with antiaircraft fire. The places where the returning planes sustained damage were correctly inferred to require no further strengthening or other improvements. Parts of the fuselage where there was no damage, on the other hand, did require attention, for if any damage had been done to these parts, it would have been fatal to the pilot.

This example illustrates how functional inferences can be made by the study of where unsuccessful attacks are distributed. Whether the structure is an airplane or a molluscan shell, the result of such an analysis is a better understanding of how the structure works and of which attributes enable it to withstand the forces to which it is exposed.

With this perspective in mind, I began systematic sampling of populations of living molluscs in order to obtain estimates of the frequency of repaired breaks sustained during the life of the builders. These repairs are recorded on the exterior shell surface as scars that cross previous growth lines. Usually they represent the trace of the outer lip when it was damaged at a previous growth stage. Some damage is not recorded in this way. Predators may, for example, break off the tips of spines or abrade the shell surface in ways that are indistinguishable from everyday wear and tear. The frequency of repair is therefore a conservative indicator of the frequency of failure by predators and other agents to break shells.

How well does a shell function as a fortress? For a given population, a first approximation to the effectiveness of the shell as armor can be obtained by calculating a simple ratio, the number of repaired breaks in living and "dead" shells divided by the total number of breaks (lethal as well as repaired) in the sample. This estimate may be too low if many of the "lethal" breaks occurred after the death of the original builder. Hermit crabs living in snail shells, for example, are susceptible to the same shell-breaking predators that prey on living snails. Damage to shells occupied by hermit crabs or to empty shells should thus not be counted in calculating the effectiveness of the shell as protection for the mollusc that built it. The postmortem effect can be evaluated owing to the lucky circumstance that a second form of predation, drilling by snails, affects only living molluscs and not secondary shell-dwellers like hermit crabs. This means that "lethal" breakage in a drilled shell must have been inflicted after the snail's death. At sites where drilling is a sufficiently common cause of death to permit estimates of the frequency of postmortem breakage, the latter accounts for less than 10% of the total frequency of "lethal" breakage. If sampling captures most instances of lethal and nonlethal breakage, therefore, field measures of the effectiveness of breakage resistance can provide a reasonable indication of the potential evolutionary role of shell breakage in molluscs.

A similar kind of analysis can be performed to ascertain how well shells work against drilling. Death by drilling is recorded by a round hole through the shell wall or, in the case of marginal drilling in clams, at the edge of the valves. A failed drilling attempt is indicated by an incomplete hole, that is, one that does not extend to the inner shell surface. The average effectiveness of the shell as protection against drilling is then calculated as the number of incomplete holes divided by the total number of holes (complete as well as incomplete). In this case, a postmortem artifact does not exist, although of course the possibility always remains that sampling does not faithfully reflect cases of lethal drilling if drilled shells are unusually prone to breakage or burial.

The study of predation and of adaptation to it is therefore an analysis of success and failure. Neither side in an encounter is perfectly adapted, which means that there is room for improvement in both the predator and the prey. For predators, the whole predation process from prey detection to pursuit to subjugation is fraught with uncertainties and risks. They may be attacked by their own enemies or be unable to prevent competitors from stealing the prey. Attributes that increase food intake, improve the ability to find and recognize prey, and shorten the time to complete the entire sequence of steps in an attack should be favored. Speed is especially important if the predators are themselves at high risk, unless the process can be accomplished from a concealed position or predators are well endowed with protective devices. For the prey, attributes that reduce the likelihood of being found, recognized, caught, and killed are potentially favored.

The evolution of improvements depends on the benefits of success and the costs of failure. If both are high, improvement is likely; if both are low, mistakes can be tolerated, and there is little evolutionary advantage for traits that improve performance. Naturalists interested in understanding how predators and their prey evolve must therefore be able to distinguish successes from failures and probe the benefits and costs involved.

Central to this inquiry is the study of how predators obtain their food. The diversity of methods predators use is extraordinary. Molluscs in their shells may be swallowed whole, enveloped and suffocated, trapped, thrown to the ground from the air, cut open, hammered, crushed, drilled, speared, poisoned, parasitized, nibbled, or killed by forced entry. Before any of these subjugation techniques can be put to use, however, the predator must first detect, recognize, and catch its victim. In the next several sections, I shall examine the predation process from beginning to end, and highlight variations on the many techniques employed by predators to kill molluscs. The chapter closes with a discussion of how predation and its attendant risks vary geographically.

Finding the Victim

For predators in the wild, the first critical step in the acquisition of food is detecting and recognizing prey. Given that prey are often rare and that many would-be victims are well adapted to avoid detection, this is a major challenge.

Some predators wait for prey to come to them. It is my impression that few predators of molluscs belong to this category. Among those that do, sea anemones lie in wait in crevices. When a dislodged snail or mussel lands in a crevice inhabited by an anemone, it will be grasped by the predator's tentacles and ushered into the digestive system.

Entrapment is also used by some nonpredatory enemies. When predatory snails linger too long near a mussel, the latter can secrete byssal threads that ensnare the snails and prevent their escape. In this position, the snails are easy prey for other animals, and in the event they are not eaten they will starve for lack of food. Some limpets are said to clamp their shells over the foot of predatory snails.

Most predators go in search of food. For them, the ability to detect visual, chemical, or mechanical cues at a distance is essential. Predators should therefore have well-developed senses. Vision probably is the primary means by which many fishes, birds, mammals, octopuses, and some crustaceans such as mantis shrimps (stomatopods) find prey. It works well as long as habitats are exposed to light or, in the case of bioluminescent animals, the predators themselves supply the light. Shallow clear-water habitats such as reefs, sand flats, rocky shores, and lakes are ideal for the use of vision; turbid rivers, mud flats, and deep-water environments are not.

Most predators of molluscs rely on chemical or mechanical receptors for the detection of food. Chemical cues released by the prey are readily detected by the sense organs in the proboscis of predatory snails and by the receptors of echinoderms. Crabs have receptors in their antennae and in the hairs on their claws; prey can be detected at close quarters by tactile means. Predators relying on mechanical receptors can detect vibrations and other changes in pressure caused by the movements of prey.

Curiously, long-distance chemical detection is virtually unknown in predators of freshwater molluscs. Sea stars and predatory snails, which are the preeminent chemical detectors of marine molluscs, are absent in fresh water. I think this absence is related to the prohibitive cost of using soluble substances to detect chemical signals and to the great benefit of preventing release of such cues. Fresh water differs from sea water in that the concentration of soluble substances is low. Yet, soluble materials such as salts and proteins are essential to the biochemical machinery of organisms in fresh as well as in salt water. In order to retain these substances in the body, they cannot be placed in situations where they will dissolve readily into the surrounding medium. If detection of chemicals is contingent on the reversible binding of these substances with receptors in the predator, some loss of soluble ions that are involved in this binding will take place in the predator. Moreover, the prey may not release many soluble cues. In this connection I have noticed that, whereas many plants and animals on land and in the sea have a distinctive odor, this is generally not the case in freshwater species.

Pursuit

Because many prey molluscs are mobile, most predators must pursue their victims once the latter have been recognized as desirable. Pursuit takes the form of crawling, swimming, leaping, or burrowing.

Among crustaceans and vertebrates, species that prey heavily on molluscs tend to be relatively slow. For example, the crabs, mantis shrimps, pufferfishes, and turtles that employ shell breakage as the chief method for subduing molluscan prey have heavy feeding devices compared to those of species that go after faster prey such as squid, shrimps, and fish. Because of this burden, these shell-breakers move slowly. The predators themselves are consequently unable to escape from their own enemies, and must therefore rely on such defenses as spines and plate armor that further reduce locomotor performance.

The situation is quite the opposite in predatory snails. Moon snails (Naticidae) and olive shells (Olividae) are some of the fastest burrowers among snails, and can overtake many of their molluscan victims. Snails envelop, enter, or drill their prey, and therefore do not possess cumbersome feeding devices that interfere with rapid locomotion. In fact, a large foot useful in enveloping a clam or snail is also ideal for high speed.

Capture and Subjugation

Predators have evolved a wide variety of techniques to extract the flesh of molluscs from the shell. These methods can be grouped into four broad categories: whole-animal ingestion, invasion by way of the aperture, breakage, and drilling.

Variations on the theme of whole-animal ingestion occur in many predatory groups. The prey is swallowed, enveloped, or smothered without damage to the shell; the flesh is then slowly digested before the empty shell is expelled. In most instances, prey size is small relative to predator size, for the predator's digestive tract must accommodate the prey's cumbersome skeleton as well as the edible flesh. Most predators employing whole-animal ingestion—many sea stars, fishes, birds, sea anemones, leeches, and segmented worms—have broad diets that include many animals besides molluscs. Only some snails, such as volutes (Volutidae), olives, and some bubble shells (opisthobranchs) have specialized molluscan diets.

Breakage sometimes occurs after the prey has been swallowed. Many wrasses (labrid fishes), as well as freshwater minnows (Cyprinidae) and cichlids, employ greatly thickened pharyngeal bones in the throat region to crack shells after the latter have been ingested. The stomachs of wrasses such as *Coris aygula* in the tropical Pacific are often packed with shell fragments, opercula, and the remains of hermit-crab skeletons that passed through the pharyngeal mill after the prey had been plucked from the sea bottom with protruding incisor-like jaw teeth. Shell destruction after swallowing also occurs in the gizzards of ducks and in a few opisthobranchs.

Whole-animal digestion is an ecologically widespread form of predation, being especially common in sandy and muddy marine habitats and in fresh water. It may be the principal mode of predation in the deep sea. The only form of whole-animal ingestion that is common on reefs involves post-ingestive shell cracking.

A second category of predatory methods is invasion. The predator enters the prey's shell by way of the aperture or, in the case of clams, between the valves. The shell is left intact while the flesh is removed or digested in place. Because food preparation takes place mainly outside the confines of the predator's digestive tract, predators using variations on the invasion method consume relatively large prey.

Many invading predators have specialized molluscan diets. Predatory snails typically insert the proboscis into the victim's shell, often poisoning or anesthetizing the victim before digestion begins. Some terrestrial pulmonate snails insert a substantial part of the body into the prey shell. Sea stars are able to extrude the stomach through an opening as small as 0.1 mm between the valves of clams, and digest the flesh while it is still in the prey's shell. Beetles, harvestmen, and some specialized snakes extend their long curved mouthparts deep inside the prey of their land-snail prey. The beaks of birds and the teeth of some fish serve a similar function.

Some invading predators employ substantial force. The northeastern Pacific sea star *Evasterias troscheli* can exert 54 N of force as it pulls apart the valves of clams with its suckered arms. Snails such as *Busycon* and *Fasciolaria* use the outer lip of their shell as a wedge to open the valves of tightly closing clams.

Time trials show that the use of force or venom significantly speeds up shell entry, which is otherwise typically a very slow method of predation. Sea stars using force to pry open clams can kill and consume prey in less than half the time required by species that do not resort to force. Cone shells, which poison their molluscan prey, take less than an hour to subdue and eat a snail; other predatory snails take many hours to consume their prey.

An interesting variation on invasion is the nibbling of exposed body parts. Flounders and many other fish often bite off the tentacles and siphon tips of clams. Some tropical wrasses and temperate dogfishes grab and bite off the exposed foot of crawling snails. Normally, this kind of predation does not result in death, but the regeneration of lost body parts does exact a toll in lost feeding capacity and reproductive success in affected individuals.

As a generally time-consuming method, predation by invasion would be most feasible either for predators that are themselves extremely well defended against enemies or for those living in relatively safe environments. The prevalence of apertural entry by predators of ground-dwelling land snails is consistent with the second of these possibilities. So is the distribution of shell-prying sea stars. These slow predators are almost entirely temperate in distribution, and are absent on most tropical shores where faster predators predominate. Most gastropods that practice invasion by way of the aperture are found on marine sandy or muddy bottoms. The chief exceptions are ranellids and some muricids, whose heavily armored shells may offer sufficient safety on the tropical rocky bottoms they typically inhabit.

Shell breakage before swallowing enables predators to take large prey rapidly. Specialized instruments—claws, jaws, and beaks—are required to crush, peel, or hammer heavily armored prey; but thin-shelled prey may be dispatched even with relatively unspecialized devices. Crabs and clawed homarid lobsters typically crush prey shells between the apposing surfaces of the fingers of the claws or chelae, the first pair of legs. These surfaces typically bear thick molarlike processes near the pivot between the fingers, where the mechanical advantage of the claw is highest (fig. 5.2). Forces up to 800 N have been measured by my former student Jay Blundon in the Floridian stone crab *Menippe mercenarla*. It is likely that even greater forces are exerted by such tropical crabs as *Carpilius*, *Eriphia*, and *Daldorfia*, in which the mechanical advantage is even higher than it is in *Menippe*. Calappid box crabs are specialized for peeling. By repeatedly breaking away pieces of the shell's lip, with a peglike tooth near the base of the movable finger of the right claw (fig. 5.3), box crabs cut a spiral incision into the body whorl of a snail shell until the soft parts are laid bare or until the crab can no longer insert the tooth into the ever smaller opening. Clawless spiny lobsters also peel shells, but they ac-

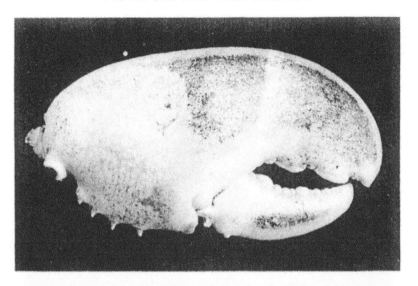

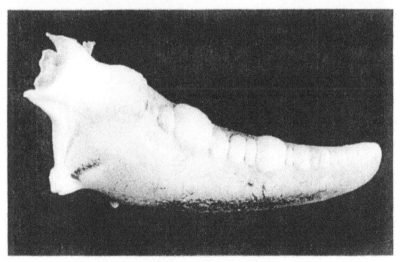

Fig. 5.2. Crusher claw of American lobster (*Homarus americanus*). Note the large molarlike teeth close to the claw's pivot, and the smaller thickenings toward the tips of the fingers. Top, Entire claw; bottom, movable finger or dactyl. The claw is 21.8 cm long.

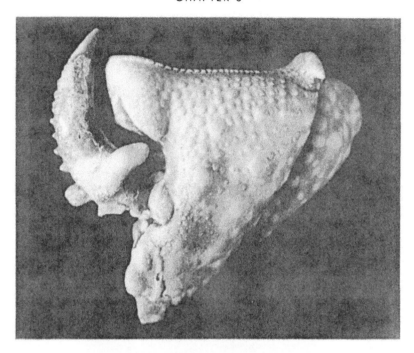

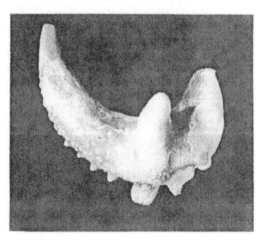

Fig. 5.3. Specialized right peeling claw of the box crab *Calappa convexa* from Panama (top). A long tooth on the outer face of the movable finger (bottom) breaks off pieces of the outer shell wall of prey snails. Top, Front surface of the articulated right claw. The height of the claw is 43 mm.

complish it with greatly enlarged right and left mandibles, whose molarlike surfaces break off successive pieces of the shell beginning at the outer lip. Many turtles, mammals (sea otters and desmans, for example), lizards, crocodiles, fishes (puffers, trunkfishes, porgies, triggerfishes, rays, freshwater drums and killifishes, and some sharks), and some beetles, harvestmen, and even opisthobranch snails use jaws for crushing. The spiny puffer *Diodon hystrix* (fig. 5.4; see also plate 8) can crush shells of the Panamanian muricid *Vasula melones* that in compression tests fail under loads of 5000 N or more. Pounding is done by certain mantis shrimps (stomatopods) with the aid of greatly enlarged limbs known as the second pair of maxillipeds. Thrushes and mongooses hammer or hurl land-snail prey against hard surfaces. Some muricids whose outer shell lip bears a downwardly directed spine fracture barnacles and perhaps other hard-shelled prey by striking the victim with the spine. Finally, gulls and crows often drop prey shells from a height of several meters, shattering the shells on rocks, roads, or parking lots.

Although breakage is ecologically and geographically widespread, it reaches its greatest specialization in the marine tropics. Many predators living on rocky bottoms break shells; so do some sand-dwellers, such as calappid crabs and many fish. Shell-breaking equipment among temperate marine predators tends to be much less specialized than is that of tropical forms; the same is generally true for most freshwater predators. Only in Lake Tanganyika in East Africa, which is home to an extraordinary assortment of heavily armored gastropods, does one find freshwater crabs whose claws are as robust as those of specialized tropical marine shell-crushers.

Among tropical marine crabs, several genera display a geographic pattern in which species from the Indo-West Pacific region (the western Pacific and Indian oceans) have larger and more robust claws than do species in the eastern Pacific, which in turn have more robust equipment than species in the Atlantic. This pattern is evident in the genera *Carpilius*, *Eriphia*, and *Ozius*, all of which are known to rely at least in part on molluscs and hermit crabs as prey. Whether size and robustness of the claw translate into greater crushing force and pressure on shells is unfortunately not known.

Drilling is a specialized form of predation in which a hole is made either through the shell wall or, in the case of some bivalve prey, at the valve edge (figs. 5.5, 5.6). Naticid and muricid gastropods use an accessory boring organ, which secretes an enzyme (carbonic anhydrase) that softens the shell, in conjunction with the tonguelike radula to

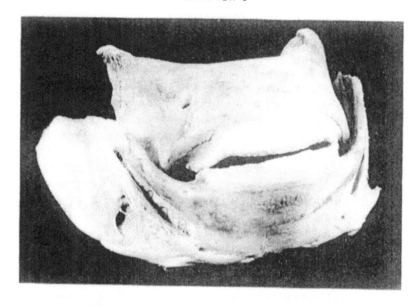

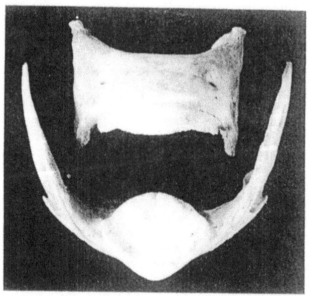

Fig. 5.4. Viselike jaws of the spiny puffer or porcupinefish, *Diodon hystrix*, from Guam. Top, Articulated jaws; bottom, separated jaws.

Fig. 5.5. Complete (right) and incomplete (left) drill holes made through valves of *Mactra chinensis* (Mactridae), Abashiri, Hokkaido, Japan. The most likely perpetrator is the moon snail *Cryptonatica janthostoma* (Naticidae). This species has an average length of 7 cm.

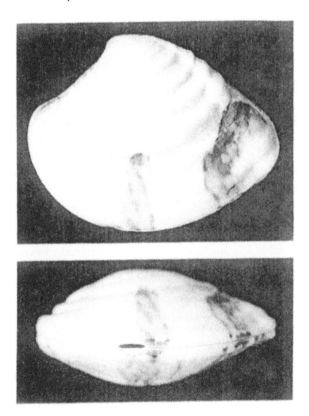

Fig. 5.6. Edge-drilled *Chione (Iliochione) subrugosa* (Veneridae), Playa Venado, Panama. The likely culprit is the moon snail *Polinices uber* (Naticidae). Top, Lateral view of left valve; bottom, view of ventral margin with drill hole. The average length of this species is 4 cm.

excavate a circular hole through which the predator's proboscis is then inserted. Octopod cephalopods drill by means of the salivary papilla to produce a more irregularly shaped hole. Some land snails may also dissolve or scrape away the shell of their victims.

Like invasion, drilling is slow. Muricids and naticids take anywhere from 5 hours to 4 days to drill a snail or clam, depending on the size and species of prey. Drilling at the valve edges appears to be faster than drilling through the shell wall. Octopods drill at least five times faster than gastropods, but it still takes 30 minutes or more for *Octopus vulgaris* in the Bahamas to drill its snail prey.

Unlike other forms of predation on molluscs, drilling is unknown in fresh water and is probably rare on land. The reasons for this rarity escape me.

The Geography of Death

For most causes of death, careful observations, monitoring, and experiments are necessary to establish their contribution to overall mortality. Because breakage and drilling leave distinctive signatures on the shells of individuals killed by these agencies, they are the only causes of death that can be inferred after the fact with some confidence.

Surveys of "dead" shells generally confirm the impression that breakage is most important as a cause of death in tropical molluscs, especially those in the western Pacific. The incidence of breakage among shells of shallow-water snails is usually between 10% and 20% in the tropics, with the western Pacific having the highest and the Atlantic having the lowest values. Of the 68 common shallow-water snail species I surveyed in the tropical western Pacific, about 20% had one or more populations in which breakage accounted for 50% or more of the mortality. Small conchs (Strombidae), horn shells (Cerithiidae), and moon snails (Naticidae) are especially prone to breakage. In a month-long study of a population of *Strombus gibberulus* at Pago Bay, Guam, for example, my wife and I found that breakage was the likely cause of death for 96% of the individuals in it.

Like many other ecological phenomena, breakage varies greatly from place to place even within a single geographic region. As a result, data from only a few sites can be quite misleading. For example, shell breakage generally kills fewer than 10% of individual snails on the cool-temperate shores of the North Pacific. In the vicinity of Akkeshi, on the southeastern coast of the northern Japanese island of Hokkaido, however, we found that no less than 42% of "dead" shells were broken. Such a high incidence would seem to be far more typical of the marine tropics, but it cannot be excluded that similarly high values will be uncovered at other temperate sites.

Just how difficult it can be to detect geographic patterns in predation was brought home to me during the course of my work with drilling. Early in my studies, I concluded that drilling was more frequent in tropical and warm-temperate prey than in species from colder seas. This conclusion was based on my study with Bettina Dudley of turritellid snails, in which the incidence of drilling by moon snails was, on the average, three times higher in the tropics than in the cool-temperate zone. When we began to examine bivalves, however, the opposite pattern seemed to emerge. The average tropical clam species in our survey showed an incidence of 9% drilling, whereas the average cool-temperate species had a drilling incidence of 24%. If we removed the cool-temperate Japanese samples, in which drilling frequencies were extremely high, the latitudinal pattern disappeared altogether. Work on shallow-water snails other than turritellids further revealed that drilling is important as a cause of death for a minority of species. Given the great variation in the importance of drilling from place to place as well as among species at single sites, a latitudinal pattern would be difficult to prove even if more extensive surveys were in hand.

The only kind of drilling for which a geographical pattern almost certainly can be demonstrated is edge-drilling. Naticids and muricids that drill clams at the valve edge are found almost exclusively in tropical and warm-temperate waters. Our surveys of empty clam valves confirm that edge-drilling is rare in cold-temperate species but is locally very common in tropical ones. In one population of the venerid *Chione subrugosa* (fig. 5.6) from Pacific Panama, every pair of empty valves displayed a neat round hole piercing both valves at the commissure. The perpetrator in this instance was probably a species of the naticid genus *Polinices*.

That edge-drilling should be common in warm water makes sense in view of the fact that this method is faster than the more typical method of drilling a hole through the shell wall. If drilling exposes the predator to risks of predation or inclement physical conditions, drilling at the commissure of bivalves or between the plates of barnacles should reduce these risks.

It is a puzzle why drilling, which is in general a very slow technique of predation, should be so common on many tropical shores. Perhaps the predators are extraordinarily well endowed with defenses, or the act of drilling is accomplished while the predator is concealed or inconspicuous. Many muricids do indeed have massive shells that provide security against shell-crushers, and naticid moon snails typically attack their victims under the sand where their activities are not easily observed by other animals.

A Geographic Perspective on Risk

Most temperate naturalists who are confronted with tropical biology for the first time come away with the conviction that life in warm climates goes on in spite of a bewildering array of dangers and risks. Animals seem to be faster, stronger, more ornately colored, more poisonous, and altogether more exuberant in their diversity than in the tamer temperate zones. Certainly the rapid methods of predation account for more deaths in tropical molluscs than in temperate ones.

Why should this be so? A partial answer is, I believe, that most biological processes operate more quickly at higher temperatures. As a general rule, there is a doubling in the rate of unregulated chemical reaction for every 10°C rise in temperature. If, for example, the rate of contraction of muscles depends on temperature in this way, then functions such as locomotion, withdrawal into the shell, and feeding will all be much faster under warmer conditions. The viscosity of water is about 45% lower at 20°C than in liquid water close to the freezing point. This, too, means that mobility and filtering require less effort at high temperatures than in the cold.

Most biological functions are regulated to some degree by enzymes. Over the range of temperature commonly encountered by individuals, rates of reaction either remain constant or rise only slightly as the temperature rises. However, this so-called acclimation is rarely complete. There is almost always some temperature-dependence, particularly near the thermal extremes. Thus, although it is true that cold-adapted scallops and fish can swim faster than would be expected on the basis of the doubling rule, their swimming is still slow compared to that of most adept tropical swimmers.

These relationships have three far-reaching consequences for individual organisms. The first is that there will be more encounters between individuals and their mobile enemies in warm than in cold environments. Individuals are therefore at much greater risk of being discovered by faster and often more potent enemies when the temperature is higher. Second, the potential for the evolution of traits useful in coping with enemies may also be greater at higher temperatures. If the number of encounters with enemies is large, individuals will be tested frequently, and enemies will play a large role as arbiters of survival. The third important consequence is that the range of adaptive options is much greater at higher temperatures. If very rapid movement and great strength are theoretically possible in cold conditions, their price in terms of other energy-consuming functions is very high. In warmer circumstances, incompatibilities are less severe and the degree to which specialization can occur in any given direction is less constrained.

An obvious way to get around these temperature restrictions is to have the body operate at a high temperature. This is, of course, what birds and mammals do. Some of us can live in the cold while maintaining all biological functions at high rates. The limitations of life at low temperatures are thus for the most part overcome, assuming that resources are available in sufficient quantity to keep metabolism running at a high level. Cephalopods, scallops, and even some snails lead active lives and have higher energy requirements than do other molluscs. The limitations on specialization imposed by a low energy budget are thus less severe for these animals.

Temperature is, of course, not the only factor that regulates the rate at which biological processes occur. The availability of food is clearly also important. Andrew Clarke at Cambridge University has, in fact, made the case that slow swimming and slow growth of most polar organisms are caused less by low temperatures than by the fact that nutrients are hard to come by for much of the year. In other words, marine animals in polar regions cannot maintain high metabolic rates because such rates cannot be sustained in the absence of food, unless they are able to migrate to other sites when food is scarce. The same problem plagues deep-sea animals, which depend for all their food on a slow rain of organic matter from the waters above. Molluscs on the upper shore and those in a regime of constant surf are also severely limited in the time available for feeding.

I know of no data on encounter rates of molluscs with their enemies as a function of temperature and food availability, but we do know that the frequencies of repaired injury of shells generally increase from high latitudes to the tropics. In other words, the number of encounters during which the shell effectively protects its maker from possible death is greater for the average tropical individual shell-bearer than for its temperate counterpart. On the cold shores of the Aleutian Islands of Alaska, for example, shell-breaking crabs are very rare, and most snail species show frequencies of repaired shells of less than 5%. By contrast, many tropical snails have an average of more than one repaired injury per shell. In one specimen of the auger shell *Terebra gouldi* from Hawaii, I counted 13 repair scars.

Incomplete drill holes are generally much less common than are repaired breaks, presumably because the chance that a drilling predator will finish its meal once it has caught the prey is very good. The highest frequencies of incomplete drilling I have seen occur in temperate clams. I have in my collection one specimen of the mactrid bivalve *Pseudocardium sachalinense* from a beach near Akkeshi, Hokkaido, in which there are one complete and four incomplete drill holes. There is no clear geographical pattern in incomplete drilling, however. All we can say at present is that incomplete drill-holes are common where death by drilling is also frequent.

If risks and the potential for evolution of antipredatory characteristics generally increase from high to low latitudes, we should expect to see a more varied array and a greater specialization of defenses in tropical as compared to temperate and polar molluscs. I consider how shells work against predators in the next chapter.

CHAPTER 6

Coping with Enemies: The Shell as Protection

The primary function of the molluscan shell is protection. Just as there are innumerable ways in which predators have evolved to find, catch, and subdue their molluscan victims, so molluscs have responded with a diverse array of evolutionary solutions. Some avoid detection by enemies or sense danger at a distance; others rely on escape or, if they are caught, on various forms of resistance. Still others occupy safe environments where enemies are rare or where the economic burden of defense falls on the organisms with which the molluscs are intimately associated. All these responses can be inferred from the sizes, shapes, and textures of shells. Much of molluscan shell architecture can be understood in terms of enemy-related adaptation. The extent and nature of the response vary according to geography and habitat, just as do the abundances and capabilities of the enemies of molluscs. Functional specialization is therefore a reflection of the evolutionary influence of enemies.

In this chapter, I go through the predatory process, from detection to pursuit to subjugation, from the molluscan prey's point of view. With an occasional aside about other functions, I examine how shells work against enemies, and how functional specialization varies from place to place.

Detection and Recognition

The first line of defense against an enemy occurs at the stage where prey detects predator and predator detects prey. This is a battle of the senses. Either party can develop senses for early and distant detection of the other, and both can prevent the other from being detected.

Shells are often specialized to accommodate sense organs. Many snails, for example, have the anterior end of the aperture modified into an extension or notch (the siphonal canal or siphonal notch) that houses the proboscis, an organ used in feeding. The proboscis is equipped with well-developed chemical receptors and sometimes with eyes. In conchs (Strombidae), a notch behind the siphonal canal on the outer lip of the aperture protects an eye that functions even when the foot of the snail is partially retracted into the shell. Many olives (Olividae) and the seraphsid *Terebellum* have a notch at the posterior end of the aperture through which receptors capable of detecting danger from behind or above protrude.

Many shell attributes make it difficult for predators to locate their molluscan prey. In well-lit environments, cryptic coloration enables some snails to blend in visually with their surroundings. The North Atlantic periwinkle *Littorina obtusata*, for example, has the shape, size, texture, and color of the bladders of its seaweed host. The snail thus receives a measure of protection from predation by small fish. In other instances, crypsis is of a chemical nature. The small Californian limpet *Tectura paleacea* incorporates chemicals from its sea-grass host into the shell. As a result, predatory sea stars are unable to distinguish prey limpets from the grass on which the limpets live and feed.

Visual predators often develop search images for prey of a particular color or pattern. Some molluscs have responded by evolving shell-color patterns that vary from individual to individual. In Kenya, for example, D. A. Smith showed that the ghost crab *Ocypode ceratophthalma* develops a search image for the most conspicuous color form of the beach clam *Donax faba* when that prey species is abundant. Individuals that are more cryptically colored thus have an antipredatory advantage. When the prey is scarce, however, the crab evidently chooses the most common color form of *D. faba*. In this case, the advantage shifts to individuals with unusual colors, perhaps including those that are easy to see. Shallow-water olives, nerites (Neritidae), and some periwinkles (Littorinidae), as well as some tropical land snails, also display great color variation among individuals and may similarly be responding to predator search images that vary according to prey abundance.

As structures that impede the passage of water and other substances, shells provide effective means of preventing detection by predators that rely on chemical cues. Such cues can be prevented from diffusing into the environment when the body is tightly sealed in or beneath the shell. A chiton or limpet, for example, can clamp its shell onto a rock to which the animal's foot clings. Clams can shut their valves, and coiled snails and cephalopods withdraw into the shell behind a tightly fitting door. If the seal is broken, the protection the shell affords against the keen chemical sense of predators is compromised. Brian Morton at the University of Hong Kong, for example, has shown that the predatory melongenid snail *Hemifusus tuba* cannot detect or recognize tightly shut clams, but readily approaches individuals whose valves have been damaged at the margin so that diagnostic cues leak out even when the shell is closed.

In fresh water, where predators using long-distance chemical detection are rare or absent, a tight seal may be much less useful than in the sea. In fact, some large unionid clams such as North American species of *Lampsilis* possess flapping extensions of the mantle that visually attract fish. When a fish comes near, it can be infected with the clam's glochidium larva, which spends its life on the fish before settling to the bottom to become a filter-feeding adult. This kind of ostentatious behavior would, I think, be quite unacceptable in the sea unless it were accompanied by appropriate defenses such as toxicity.

A hermetic seal cannot be maintained for long unless energy requirements can be reduced significantly. While the soft tissues are cut off from the outside world, the animal cannot feed, respire, or move. The price of preventing detection by retreating into a closed vessel is therefore the curtailment of other life functions.

The ability to seal the body into the shell is important not only in avoiding detection, but also as a defense against predators after the prey has been caught. I defer a discussion of these benefits until later in this chapter.

Silent Movement and Rapid Escape

Another way in which molluscs can be detected by their enemies is through movement. As an animal moves through water or sediment, it disturbs the surrounding medium. Predators can often pick up this noise with receptors that detect vibrations and other changes in pressure. It is therefore useful for potential victims to reduce noise as they travel. This can be accomplished by streamlining the body and by other means of reducing drag, especially pressure drag. The benefits of remaining undetected by enemies may be a much more important consequence of streamlining than are the energy savings that a reduction in drag makes possible.

Escape is the next antipredatory option in an encounter with an enemy after one party has detected the other. Movement to a safe place can be very effective if a relatively slow predator is in pursuit or if the prey has early warning of a predator's presence. Sometimes, escape involves little more than falling off a rock or sinking passively and noiselessly through the water to the bottom; but for many molluscs, escape is an energetically costly affair that relies on speed, acceleration, maneuverability, and endurance. Escape responses have been especially well documented in snails and clams fleeing from predatory gastropods and sea stars. Swimming and leaping are particularly effective means of escape. A few examples are known in which passive or active flight works against crabs and fishes. Even humans can elicit escape responses in molluscs. My footsteps on shores in Madagascar caused hundreds of individuals of *Nerita doreyana* to lose their grip and clatter into crevices all around me as I walked.

I have already pointed out in chapter 4 that shell attributes indicating high locomotor performance are more common and better developed in tropical than in temperate and polar marine molluscs. Freshwater molluscs are conspicuously poor burrowers and swimmers, and show none of the specializations to rapid or sustained locomotion that many marine species do.

For the most part, however, predators such as crabs, fish, birds, octopuses, and insects are too fast for the modest locomotor capacities of even the fleetest shell-bearers. Accordingly, molluscs tend to rely heavily on defenses whose effect is to slow down or deter attempts by the predator to subdue its victim.

Interfering with Manipulation

The first thing a predator must do after it has caught its victim is to manipulate it so that subjugation can commence. Shells must be enveloped in the foot of a snail for suffocation or drilling, picked up and manipulated by the claws of crabs, maneuvered into a fish's mouth, held by the suckers of a leech or a sea star, and so on. Various attributes of shells function to make this initial step of subjugation more difficult.

Large size provides a very widespread deterrent to manipulation by predators. Almost every study of predation shows that the predator's chance of success drops off as prey size increases. I know of only two exceptions to the general advantage that large size provides. When gulls and crows drop molluscan prey on a hard surface, the force with which the shell strikes the ground is gravity, which scales with the third power of the prey's linear dimension, that is, it is proportional to mass. Shell strength, however, tends to be proportional to the square of shell-wall thickness, which is the second power of a linear dimension. A smaller shell is thus less at risk of shattering than is a large one, even if it is easier to carry aloft. Whether this exception is significant depends on where the prey is dropped. If marine molluscs are released on a high cliff or parking lot, they will die of desiccation if they do not shatter. Snails dropped on a stony beach might survive if their shells remain intact.

The second exception occurs in high-spired snails under attack by calappid box crabs. The aperture of young individuals is often so small that an adult crab cannot maneuver the specialized tooth of its right claw into the opening to begin peeling. As a result, the prey is rejected unharmed. Larger snails, however, have an aperture big enough for the tooth to be inserted properly.

In many molluscs, large effective size is achieved by the development of high-relief sculpture. The functional benefits of knobs in making shells larger and therefore more difficult to swallow were nicely demonstrated by A. Richard Palmer in experiments with the predatory spiny pufferfish (*Diodon hystrix*) in Panama. He offered the predator a range of sizes of the muricid snail *Thaisella kiosquiformis* (fig. 6.1). One group of prey had the large protruding knobs filed off, whereas the other was offered with the sculpture intact. The upper size limit at which the prey could be successfully crushed in the predator's powerful jaws was about 5 mm greater for the altered snails than for the unaltered controls. Because the maximum gape between the puffer's upper and lower jaws determines whether a shell will fit into the mouth, projections that increase a snail's linear dimension between the dorsal and ventral surfaces of the shell protect a snail from crushing even if the prey's shell is thin.

Nothing about shells evokes the tropics more for me than do knobs and spines (figs. 6.2, 6.3). These features occur in 20%–35% of the snail species that live on hard surfaces in the tropical Pacific. The incidence is a little lower (10%–18%) in the tropical Atlantic, and much lower (less than 2%) in the cool-temperate zones. Sand-dwellers generally lack knobs and spines, presumably because they interfere with burrowing; however, many conchs (Strombidae), helmet shells (Cassidae), and Indo-Pacific mud snails (Nassariidae) have a conspicuous knoblike varix (fig. 6.4; plates 9–11) located dorsally on the adult shell. In fresh water, highly sculptured species are few and far between. Good examples occur among tropical species of *Clithon* (Neritidae) and *Brotia* (Thiaridae), as well as in the endangered pleurocerid *Io spinosa*, found in the rivers of Virginia and Tennessee in the southeastern United States. Several of the unusual snails in Lake Tanganyika are also highly sculptured. Spines and knobs are, as far as I know, not present in land snails.

A very different tack is taken by cowries and other molluscs with highly polished slippery shells (fig. 6.5). In Guam, we have often seen crabs try to manipulate cowries only to have the prey catapult across the aquarium floor as the predators lose their grip. In cowries (plates 12, 13), marginellids, and olives (plates 6, 7), the polished surface is

Fig. 6.1. *Thaisella kiosquiformis* (Muricidae), Tivives Estuary, Pacific coast of Costa Rica. This species, which normally lives on mangrove trees where it feeds on barnacles, bears strong knoblike spines. Left, Apertural view; right, dorsal view. This species reaches an average length of 5 cm.

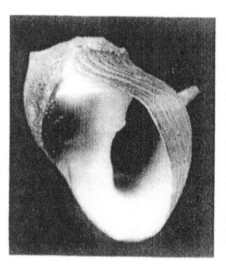

Fig. 6.2. *Clithon brevispinus* (Neritidae), Ngermid, Palau. This is one of the few freshwater snails with spines on its shell. It inhabits fast-flowing streams, where it lives on rocks. Left, Apertural view; right, dorsal view. This specimen is 15 mm in diameter.

Fig. 6.3. *Hysteroconcha lupanaria* (Veneridae), Las Lajas, Panama. This sand-dweller bears extraordinarily long spines near the posterior end of the shell. These spines may prevent the clam from being swallowed by fishes, and may also provide stability in the sediment by making the posterior end of the shell functionally truncated (see chap. 4). The shells of this species can be as long as 39 mm with spines up to 18 mm in length.

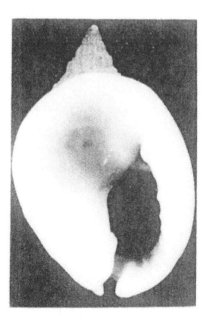

Fig. 6.4. *Nassarius (Plicarcularia) pullus,* Wom Village, Papua New Guinea. Apertural view showing thickened outer lip and broad thick ventral pad or callus; a knoblike varix is on the dorsal side. This mud snail (member of the Nassariidae) burrows in sand. This species reaches a length of 15 mm with a callus 14 mm long.

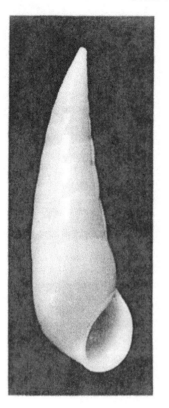 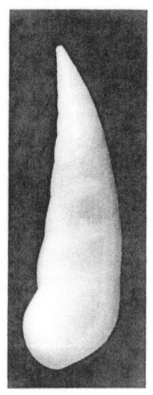

Fig. 6.5. *Melanella dufresnei* (Eulimidae), Piti Bay, Guam. This extremely slippery smooth shell belongs to a snail that feeds on sea cucumbers. When not on its host, it lives in sand. The shell's axis is conspicuously curved. If this curvature has any functional significance, it has not yet been discovered. Left, Apertural view; right, dorsal view. This specimen is 25 mm long.

deposited by the surface of the mantle or an extension of the foot, which in life usually covers the outer shell surface. The soft parts can, however, withdraw fully into the shell, so that the latter presents a slippery surface to would-be attackers. That the polish is adaptive and not simply an inevitable consequence of having been laid down by the surface of the mantle or foot is indicated by the fact that not all snails with such partially or wholly internal shells have a smooth surface. Some abalones (Haliotidae), keyhole limpets (Fissurellidae), and even a few cowries (the ovulid *Calpurnus*, for example) have matte surfaces that are not slippery to the touch, even though the shell is covered by a flap of the mantle (figs. 6.6, 6.7). Some snails (bubble shells and sand-dwelling cone shells) and razor clams (So-

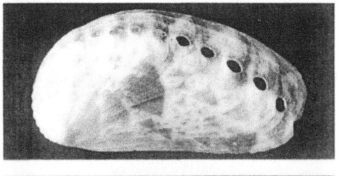

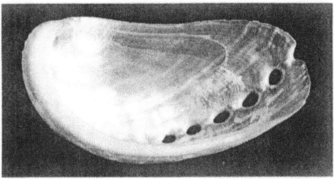

Fig. 6.6. *Haliotis asinina*, Ulang, Palau. The shell is smooth but not slippery even though it is partially enveloped in life by the mantle and foot. The row of holes seen in both views has a respiratory function; waste water is expelled through the holes. Top, Exterior view; bottom, interior view. This specimen is 46 mm wide.

Fig. 6.7. *Calpurnus verrucosus* (Ovulidae), Lolorua Island, Papua New Guinea. This cowrie relative has its shell enveloped by the mantle, yet it is not at all glossy. The animal lives and feeds on soft corals. The function of the conspicuous dorsal hump is unknown. Left, Apertural view; right, dorsal view. This specimen is 28 mm long.

lenidae) owe their slipperiness to a thin layer of periostracum. Finally, there are some snails such as pheasant shells (Phasianellidae), eulimids (fig. 6.5; plates 12, 13), and rissoids of the genus *Zebina* whose slippery surface appears to be wholly external and to lack a substantial periostracum.

Shell-Wall Thickening

Once the prey has been brought into proper position for the final assault, the shell must function as a fortress. Just as the defenders of castles in Medieval Europe relied on thick walls to keep out enemies, so molluscs have often developed extremely thick shells. Resistance to breakage generally increases as the square of wall thickness. All else being equal, a doubling in the thickness of the wall thus yields a four-fold increase in strength. The presence of tiny cracks, however, seriously weakens the wall, because forces that stress the shell are concentrated at cracks. If cracks are more frequent in a thick wall than in a thin one, wall thickening would not necessarily result in greater strength. It is probably for this reason that measurements of shell strength reveal great variation among shells of similar mass and thickness.

This variation is very important. Predators usually fail in their attempts to break a very thick shell. If thickness were a reliable guide to shell strength, predators could probably learn not to waste their efforts on thick shells. Crabs might nevertheless attack excessively thick shells just in case unobservable cracks render the fortress vulnerable. We have often seen crabs attack shells that by the usual criteria would be unbreachable. By testing such oversized shells, crabs and other predators are, as it were, keeping the pressure on their victims evolutionarily to maintain resistance defenses.

Thickening also has an effect on the time required for predators to subdue prey. A predator may eventually be able to break a thick shell by squeezing it repeatedly and so initiating tiny cracks that ultimately spread and cause the shell to shatter, but such manipulation takes time. Our crabs in Guam worked on shells of *Drupa* species for up to an hour before they were able to get at the snails' flesh. In nature, the luxury of time may not be available very often. Drilling predators are similarly affected. The time required for snails to complete a drill hole increases linearly with the thickness of the wall that must be penetrated.

Shells need not be uniformly thick to be effective fortresses. In many tropical snails, the adult lip is the thickest part of the shell (plate

13). Experiments with the sand-dwelling crab *Calappa hepatica*, which typically attacks snails first at the outer lip, show that a thickened apertural rim serves as an effective defense. Above a thickness of 1.6 mm, the adult lip of *Strombus gibberulus* (fig. 6.8; plates 14, 15) could not be breached by a *Calappa* of average adult size (a width of 70–80 mm). When the shell-crushing crab *Carpilius maculatus* attempts to break the shells of *Drupa* (plates 14, 15), whose thickened adult outer lip is beset with massive thickenings on the inner edge and by knobs and spines on the external edge, success is achieved only when the shell is cracked on its upper or left side (fig. 6.11). The lip usually remains intact even though knobs may occasionally be abraded by the crab's actions.

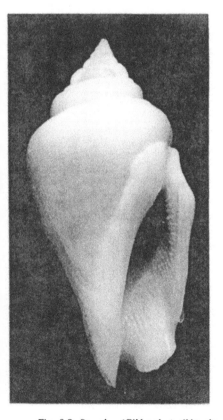 

Fig. 6.8. *Strombus (Gibberulus) gibberulus gibbosus*, Cocos Island, Guam. The narrow aperture of the adult is bordered by a thickened outer lip bearing a distinct notch toward its anterior end. One of the snail's eyes protrudes through this notch. This species reaches a length of 49 mm.

Strombus and *Drupa* have thickened lips only as adults. While the shell is still growing, these snails have thin lips that are highly susceptible to breakage. Our work on *S. gibberulus* in Guam, together with many other studies by others, shows that growth during the thin-lipped phase is extremely rapid, so that young individuals spend relatively little time in this vulnerable phase. Moreover, thin-lipped individuals tend to conceal themselves under stones, in sand, and in other sites where encounters with potent shell-breakers are rare.

Many snails have thickened lips, or varices, developed at regular intervals at the apertural margin as the shell grows. These varices can be very effective defenses if the foot can be withdrawn behind the one formed previous to the last varix. In sand-dwelling cerithiids of the genus *Rhinoclavis*, for example (fig. 4.16), attacks by *Calappa* almost invariably cause the thin last varix to break, but continued attempts by the crab to peel the shell in a spiral direction are thwarted by the varix located a little more than one-half of a whorl back from the outer lip. The snail, whose foot can withdraw beyond this varix, is subsequently able to repair the damage.

Varices probably have functions in addition to imparting strength. Many warm-water snails have varices spaced at intervals of 90° around the shell. This means that, when the outer lip is itself a varix, the preceding varix is dorsal in position. The shell thus comes to have a somewhat triangular cross-section. A shell of this form falling through water is more apt to land aperture-down than is one from which the dorsal varix has been artificially removed. Even if the snail landed upside down, the dorsal varix ensures that the aperture and foot face more or less to the side rather than up. The ability to right the shell quickly is therefore enhanced by the presence of the dorsal varix. This can be very important if snails are at risk of having part of the foot bitten off by a hungry fish.

Most snails with varices are found in the marine tropics, where they comprise 10% to 25% of the species in local assemblages. Examples may be found in many families, including the Strombidae (conchs), Cassidae (helmet shells), Ranellidae (triton shells), Muricidae (murexes), Columbellidae (dove shells), Nassariidae (mud snails), and Cerithiidae (horn shells) (figs. 6.4, 2.9, 2.10; plates 9–11). Regularly spaced varices are known in less than 2% of temperate marine snails and are wholly unknown in snails in fresh water and on land.

Another area of the shell that is often greatly thickened is the ventral part of the body whorl (fig. 6.4; plates 12, 13). There is often a thick glaze or callus on this part of the shell, which is supported from below by the snail's foot. Spectacular examples occur in conchs, hel-

met shells, murexes, mud snails, and button shells (umboniine trochids) (plates 9–11). In addition to making the shell thick where it is
most vulnerable to being broken when squeezed in a jaw or claw, the
ventral glaze also lowers the shell's center of gravity. As Enrico Savazzi
at the University of Uppsala has noted, this could be important for
snails that execute rapid movements such as leaping or lunging during escape maneuvers or while pursuing prey. The low center of gravity tends to keep the shell balanced and makes righting easier if the
animal is accidentally overturned.

Once again, the presence of a thick glaze on the underside of the
shell is a warm-water marine phenomenon that is especially well developed in sand-dwelling species in the Indo-West Pacific region. Rock-
dwelling murexes and nerites that develop this feature are also tropical in distribution. Many freshwater nerites and land snails have a
broad smooth area to the left of the aperture that is supported by the
foot, but this area is usually neither thickened nor well demarcated
from the rest of the shell.

Corrugations and Buttresses

Thickening provides an effective means of strengthening the shell,
but it does so at a potential cost. Shell material is dense. As a result,
ways of life requiring rapid locomotion or buoyancy are incompatible
with a massive shell. The incompatibility can be relieved to some degree by strengthening a relatively thin shell with corrugations and
buttresses, which add stiffness but little mass.

When a compressive force is brought to bear on the convex outer
surface of a shell, it causes the shell wall to bend ever so slightly. If the
edges of the shell are free, as in a clam valve or a limpet, they will tend
to splay out as force is applied. The outer layer of the shell undergoes
compression—that is, its structural elements are pushed together—
whereas the inner layer is under tension so that its components are
pulled apart. The middle layer (or neutral surface) is under neither
compression nor tension. If there is nothing to stop the wall from
bending, the shell is apt to break in tension before it does so in compression, because shell material is much stronger in compression.
Structural features that prevent or reduce bending therefore
strengthen the shell. These features may be classified as corrugations
or buttresses.

Corrugations, or folds in the shell wall, provide stiffness by concen-
trating shell material in the layer that is under the greatest compression when force is applied. Layers that would be in tension are not

present. Such features are very widespread in molluscan shells. All ribs, knobs, and spines are initially constructed as folds or hollow tubes at the outer lip and therefore impart stiffness to it. Many thin-shelled molluscs have a pleated shell wall. Usually these folds are spiral or radial elements, so that the greatest resistance to breakage is in the direction perpendicular to the growing margin. Such folds occur in many tusk shells (scaphopods), snails (tonnid tun shells, cold-water buccinid and neptuneid whelks, and many limpets), clams (pectinid scallops, cardiid cockles, and gryphaeid and ostreid oysters, and tridacnid giant clams), and fossil cephalopods (fig. 4.9). Concentric folds occur in some burrowing clams (fig. 4.13). Fossil cephalopods, especially ammonoids, were often corrugated by folds parallel to the outer lip. So are many thin-shelled deep-water and cold-water snails.

The cold-water snails just mentioned notwithstanding, corrugations are especially prominent in warm-water marine shells. In tropical assemblages of clams living on hard surfaces, more than 20% of species have radial folds, whereas in cool-water faunas the incidence is usually less than 15%. In fresh water, clams with scalloped posterior margins occur in various warm-water rivers. Their highest incidence is in eastern North America, where about 20% of species have these wavy edges.

In a dome, the edges of the surface that is being loaded by a force are fixed in position and therefore cannot splay out. The application of force from the outside first causes the surface to bend slightly, but as the limit of bending is reached, the force is transmitted through the wall to the supports to which the edges are affixed. The force then becomes entirely compressional. The edges of the surface can be fixed in several ways. In coiled snail shells, the shell as a whole acts as a dome, having no free edges except at the outer lip. If a force is applied from above on a shell whose aperture faces the ground, the outer lip will be mainly in compression if its surface is steeply inclined to the ground. For clams whose valves are squeezed together, the free edges will be in compression rather than in tension if the valves are highly inflated, so that the angle of contact between the two valves is large. The inner margins of the valve edges may bear riblets or crenulations that are not quite parallel to each other. When the valves are squeezed together, the riblets prevent the margins from splaying out, so that most of the force is converted to compression.

The buttressing effect in coiled snails is best developed when coiling is tight, that is, when overlap between adjacent whorls is high. Overlap means that part of the apertural rim is built on, and therefore supported by, the preceding whorl. Buttressing is incomplete in shells possessing an umbilicus, a depression on the base in which earlier whorls are exposed. Parts of the shell in the vicinity of the umbilicus are unsupported by previous coils and are therefore susceptible to catastrophic breakage.

The geographic and ecological distribution of umbilicate shells is consistent with the idea that umbilicate coiling incorporates a structural weakness and is therefore not well suited to the breakage-ridden environment of the marine tropics. Shells with an umbilicus characterize less than 5% of tropical Pacific hard-bottom snail species, 5%–10% of tropical Atlantic species, and more than 10% of northern cool-temperate species. In the tropics, most umbilicate species either live on heavily wave-exposed shores (the West Indian trochids *Agathistoma excavata* and *Cittarium pica*, for example; figs. 4.4–4.6) or as tiny snails on seaweeds. Freshwater snails show an incidence of 35% or more of umbilicate species. Land snails, too, often possess an umbilicus, but no detailed figures of the frequency of umbilici are available.

The umbilicus can be geometrically eliminated in two ways. The first involves plugging the umbilicus with a pad or callus of calcium carbonate. Such plugs are known in many moon snails (Naticidae) and in sand-burrowing button shells (trochids of the genus *Umbonium*).

The second way of eliminating the umbilicus without changing the whorl expansion rate is to build a high-spired shell. Such turreted shells are very widespread. In tropical Pacific assemblages of sand-dwelling snails, they are found in 20%–35% of species. Less than 20% of tropical Atlantic sand-dwellers have high spires, and almost no temperate species do. Hard-bottom marine snails tend to be less turreted. Only 10%–20% of species in tropical assemblages, and less than 5% of species in the temperate zones have a turreted shell. In fresh water, high spires are mainly tropical, occurring in 15%–50% of species in assemblages in Africa and Southeast Asia. On land, high-spired shells are especially characteristic of species that burrow in the ground or that live on tree trunks and cliffs.

Buttressing reaches its zenith in the shells of cephalopods. These animals are exposed to forces not only from predators, but also from the difference in pressure between the shell interior and the outside (see chapter 4). David Jacobs has argued that the shell is interpretable as a series of domes, each supported beneath by the junctions of the internal septa with the shell wall (fig. 6.9). The septa themselves are also domes whose free edges are constrained by the outer wall of the shell. Many cephalopods, especially ammonites of the Mesozoic era,

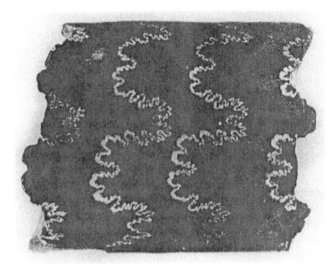

Fig. 6.9. *Baculites* sp., Mesozoic, Arrow Lakes, British Columbia. The sinuous lines mark the juncture of the chamber septa with the outer wall of the shell. Top, Lateral view of the suture line; bottom, view of the surface of the septum.

have extremely complex septa whose edges are serrated, frilled, and fluted. The contact between septum and shell wall is therefore very extensive, with the result that much of the outer wall is well buttressed from within. Such buttresses enabled some cephalopods to evolve highly streamlined shells, which are characterized by almost flat right and left faces. Such flat faces would be susceptible to being loaded in tension when a force is applied from the outside, but this force is effectively resisted by the buttresses within. All this is accomplished while keeping the shell thin and lightweight.

Reducing the Effects of Damage

Few architectural designs can prevent shells from being damaged during an attack by a determined predator, but some kinds of damage are worse than others. For many molluscs, even quite superficial damage is accompanied by high risk. Figure 6.10 dramatically shows what I mean. In the illustrated specimen of *Ficus ventricosus* from Ecuador, superficial damage to the outer lip was accompanied by a mantle in-

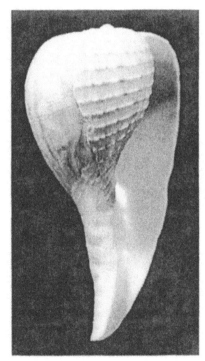

Fig. 6.10. *Ficus ventricosus*, Punta Centinela, Ecuador. Left, Normal individual, showing the widely spaced spiral corrugations. Right, Individual that was severely injured. After the shell was repaired, the spiral sculpture was much finer than the normal spiral sculpture produced before the injury. Evidently, the mantle was severely damaged at the same time the outer lip of the shell was broken. This specimen is 70 mm long.

jury so grave that the part of the shell produced by the imperfectly healed mantle has radically different sculpture from that produced by the uninjured mantle earlier in life. The soft body of *Ficus* cannot withdraw into the shell. Most shell damage is therefore probably either fatal or very costly in terms of tissue injury.

Similar problems beset most clams. As I mentioned earlier in connection with preventing detection by predators, even minor damage to the valve edges puts a clam at risk of being discovered and invaded by other enemies. This circumstance may place limits on the extent to which breakage resistance can evolve. If the elaboration of such resistance requires that individuals attacked by shell-breakers survive and reproduce, the risks associated with minor shell damage must be small.

In the case of limpets, the shell provides good protection as long as its edges fit snugly against the surface to which the animal clings. In the event of dislodgment, however, the shell is of almost no use at all. This may explain Richard Lowell's finding that the force required to dislodge a limpet is typically greater than that needed to break the shell.

Given this shortcoming, it is not surprising that marine limpetlike snails are chiefly a high-latitude phenomenon. Limpets comprise 1%–6% of shallow-water hard-bottom snail species in the western tropical Pacific, 14%–21% of species in tropical America and West Africa, and more than 25% of species in the cool-temperate North Pacific. Most tropical limpets either live in areas of strong surf, where they often occur in self-made depressions from which they are very difficult to dislodge, or on the shells of other molluscs, where they may be relatively safe from shell-breaking predators. There are relatively few freshwater limpets, comprising not more than 10% of most assemblages. Nevertheless, the limpet form has arisen several times independently.

Chitons are limpetlike in that the shell acts as a shield protecting the soft tissues and in that superficial shell damage would likely damage soft tissues as well. Unlike limpets, however, chitons can roll up when dislodged. In many species, the head valve almost touches the tail valve in the enrolled condition, and the flexible girdle surrounding the shell curls to reduce access to the exposed foot. Whether this behavior is a useful antipredatory defense is not known.

Designs in which shell damage can be sustained without risk or injury to the soft parts are highly beneficial. Long external spines, for example, can be snapped off without breaking the shell wall as a whole. Crabs like *Carpilius* are often successful in removing spines from the outer lip of *Drupa* (fig. 6.11), but in the end the snail is often rejected because other shell defenses cannot be breached. Many snails and clams have a shell microstructure that prevents cracks from traveling away from the edge of the outer lip. Damage initiated at the outer lip is kept superficial because cracks are induced to spread parallel to the growing edge.

In some clams, the edge of one valve projects slightly beyond that of the other. The projecting edge can thus sustain damage without the seal between the valves being broken. Edge-drilling predators may also be deterred by this configuration. Clams with such projecting margins include some ark shells (Arcidae), scallops (Pectinidae), basket clams (Corbulidae), and members of the deep-burrowing family Thraciidae. In some other bivalves, notably in members of the Veneridae, the outer edges of both valves project slightly beyond the line of contact between the valves when the shell is closed.

Probably the most effective form of damage control, and the one that reduces the risk of damage to vital tissues the most, is deep withdrawal of the soft parts from the rim of the aperture (fig. 6.12). Large parts of the shell near the growing margin can be broken off during an attack without injury to the tissues. Most snails and scaphopods, and a few clams, are capable of such deep withdrawal.

The ability to retract tissues back from the shell margin is associated in snails and scaphopods with a low expansion rate of the aperture, that is, with a tall, narrow, conical shape. For a crawling animal like a snail, this shape is infeasible unless the cone is coiled into a compact form. I believe that shell coiling was an early and effective way of protecting crawling molluscs from all kinds of predators and for making the shell better suited to sustain damage at the apertural rim.

Most bivalves cannot retract their tissues beyond the valve edges, but interesting exceptions occur in a number of marine families whose members live attached to rocks or are partially buried in mud and sand. In pen shells (Pinnidae), wing oysters (Pteriidae and Isognomonidae), jingle shells (Anomiidae), true oysters (Ostreidae), and some deep-sea scallops (Propeamussiidae), the marginal portions of the valves are thin and flexible and are therefore easily damaged by predators (fig. 6.13). The mantle, however, can withdraw far from the edges of the valves, and is therefore often not injured. Moreover, a broad seal is created when the flattened flexible valve margins are squeezed together as the shell is shut. As with so many other protective devices, flexible margins are mainly tropical. Between 16 and 35% of hard-bottom clam species in the tropics are characterized by flexible margins, as compared to 10% or less in temperate species.

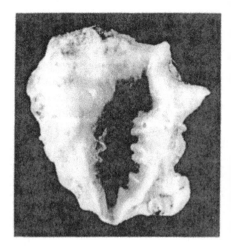

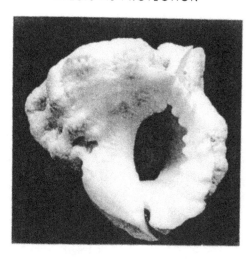

Fig. 6.11. Shells of snails from Guam broken by the crab *Carpilius maculatus.*
Opposite page: top left, *Cerithium nodulosum*; extensive lip damage. The average length of this species is 100 mm. Top right, *Trochus niloticus*; apex broken
off, some apertural damage. Shells broken by crabs reach a diameter of 38 mm.
The species may exceed a diameter of 150 mm. Bottom left, *Drupa arachnoides*;
the spiny shell has the thickened outer lip intact. This species has an average
length of 30 mm. Bottom right, *Drupa morum*; the knobbed shell has one of the
knobs damaged, but the lip was not breached. Adults attain an average length
of 40 mm. Above, *Bursa bufonia*; the thick varix at the outer lip held during the
attack, but the thin dorsal wall of the shell broke. On the average, this species
reaches a length of 60 mm. The intact lip of this specimen is 7.0 mm thick, and
the broken dorsal shell wall is 1.2 mm thick.

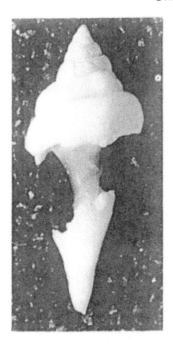

Fig. 6.12. Shells of snails from Guam broken by the crab *Calappa hepatica.* Top, *Strombus gibberulus.* Average adult length is 30–35 mm. Bottom left, *Rhinoclavis fasciata,* a high-spired shell. Average length of this species is 30 mm. Bottom right, *Polinices mammilla.* The average length is 35 mm. The shell wall of this specimen is 1.2 mm thick.

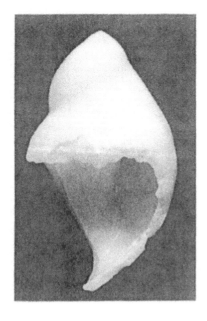

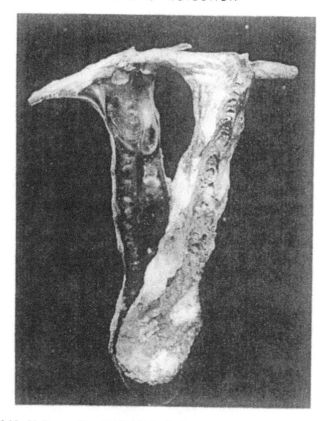

Fig. 6.13. *Malleus malleus* (Malleidae), the hammer oyster, Babeldaob, Palau. The long straight hinge, which is oriented parallel to the sand, is buried just beneath the sand surface. The rest of the shell is held above the sediment. The clam's body occupies a very small portion of the space inside the shell, demarcated by a conspicuous ridge. Outside this ridge, the shell is thin and flexible. Adults reach a height of 18 cm.

Guarding the Entrance

The aperture offers an obvious way into a prey shell for a predator. It is therefore not surprising that modification of the aperture to make it less accessible has been a recurrent theme in the evolution of shells.

Deep withdrawal, which is associated with a low expansion rate and therefore a small aperture, is a widely used and effective means of restricting access. Predators using variations on the technique of apertural invasion must have very long feeding organs to reach the edible retracted tissues of some snails. Some auger shells (Terebridae), for example, can withdraw the foot more than two whorls back from the apertural rim, and may have as many as 46 whorls (plates 16, 17).

Another effective defense of the aperture is to impede the entrance with a barrier such as a door or a series of projections extending from the rim across the opening (fig. 6.14). Snail-eating beetles and snakes are deterred by an operculum and by large folds on the outer and inner lip in many land snails. Tightly fitting opercula, like tightly closing valves in clams, enable molluscs to pass unscathed through the digestive tracts of whole-animal ingesters like some fish and sea stars.

Apertural barriers occur very widely, especially in the marine tropics. Thickenings extending from the rim across the opening are spectacularly developed in cowries, marginellids, miter shells (Mitridae), dove shells (Columbellidae), muricids, nerites, and many land snails ranging from tiny endodontids to large camaenids, streptaxids, and polygyrids (plates 14, 15). The late Alan Solem estimated that about one-quarter of the world's land snail species have such apertural barriers. Among rocky-bottom marine snails, outer-lip denticles occur in

Fig. 6.14. *Quimalea pomum* (Tonnidae), Babeldaob, Palau. The narrow aperture is occluded by teeth extending from both the outer and inner lip (left). External sculpture consists of strong spiral ribs, which in young individuals are probably corrugations (right). The species preys on sea cucumbers and lives buried in sand. This specimen is 46 mm long.

35%–40% of species in the tropical Pacific, 15%–19% of species in the tropical Atlantic, and less than 5% of cool-temperate species. Less than 25% of tropical sand-dwellers have these outer-lip denticles, but again this incidence is higher than it is in the temperate zones. Occluding folds in the inner (columellar) side of the aperture are found in more than half the snail species in most tropical assemblages from sand and mud bottoms. In those few cool-temperate sand-dwellers in which columellar folds are developed, these folds are weak and do not significantly impede entrance into the opening. Freshwater snails show no evidence of apertural restriction by folds or denticles at the lip. Some planorbids have internal plates that strengthen the shell, but these are generally not located at the rim of the opening.

Doors closing off the aperture when the soft parts are retracted are characteristic of gastropods and many fossil cephalopods. Usually, the door (operculum) of snails is a stiff organic plate carried on the dorsal surface of the foot, but in some groups it has become partially or wholly calcified. Examples of such mineralized opercula occur in the marine turban snails (Turbinidae) (fig. 6.15), pheasant shells (Tricoliidae and Phasianellidae), nerites (Neritidae), and some moon snails (Naticidae), as well as in an assortment of freshwater snails (the littorinid genus *Cremnoconchus* in India, apple snails of the warm-water family Pilidae, and members of the geographically widespread family Bithyniidae). Freshwater neritids also have calcareous opercula, but these doors tend to be thin when compared to those of most marine species. In nerites and the moon-snail genus *Glossaulax*, the operculum is hooked beneath the inner lip so that it rotates on a hinge when the snail extends or retracts its body. Even some land snails have calcareous apertural covers. In the Pomatiasidae, Cyclophoridae, and Helicinidae, these doors are true opercula, but in the pulmonate family Clausiliidae of the Old World, the covers originate as parts of the inner lip that have become separated from the rest of the shell. In cephalopods, structures closing off the aperture are modified jaws. Usually these are thin organic plates, but in some Jurassic and Cretaceous ammonoids the door is a two-part calcified structure.

Curiously, a rigid mineralized operculum is not a widely used form of apertural protection. It is more common in the tropics (5%–12% of marine species) than in the temperate zones (less than 2% of hard-bottom snail species), and within the family Turbinidae there is a trend for tropical species to have thicker opercula.

Another way to restrict entry into the shell is to make the opening narrow. Many tropical marine snails and some land snails have evolved very narrow apertures whose length may be more than ten

Fig. 6.15. *Ninella torquata*, Long Reef, near Sydney, Australia. This turbinid snail possesses a spirally coiled calcareous operculum, which fits snugly into the aperture when the soft parts are withdrawn (top, operculum). A deep umbilicus is evident on the base of the shell (bottom, apertural view). This specimen is 64 mm in diameter.

times their width. Good examples occur among marine dove shells (Columbellidae), conchs (Strombidae), miters (Mitridae and Costellariidae), turrids (Turridae), and especially cones (Conidae), as well as in the terrestrial Oleacinidae (fig. 5.1; plates 6, 7). Half of these snails have lost or greatly reduced the operculum. The incidence of narrow apertures, defined as those in which the ratio of length to width is 2.5 or higher, is 35% to 50% among hard-bottom snails in the western Pacific and Indian oceans. Elsewhere in the sea it is lower, being 15%–25% in the tropical Atlantic and less than 11% in the northern cold-temperate zone. In most tropical assemblages from shallow-water sand bottoms, 50%–60% of species have a narrow aperture. Only two genera of freshwater snails, the marginellid *Rivomarginella* of Southeast Asia and the North American lymnaeid *Acella*, could be described as having a narrow opening.

Bivalved shells can restrict entry by placing barriers across the opening between the valves. Several cockles (Cardiidae), including the tropical West American *Mexicardia procera*, (plates 18, 19) have the posterior radial ribs extended in the hind region of the shell so that, when the valves gape slightly, the ends of the ribs still interdigitate across the opening. Many oysters (Gryphaeidae and Ostreidae), scallops (Pectinidae), and all giant clams (Tridacnidae) have broad radial folds (plate 22). In some oysters, these are V-shaped corrugations. As the valves open, the straight flank of a fold in one valve moves parallel to the flank of an adjacent fold in the opposite valve. The distance between the valve edges thus remains constant as the valves open except at the peaks and troughs of the folds. This device, which Martin Rudwick first recognized in fossil bivalved brachiopods, enables the bivalve to maintain a small distance between the valves even when the angular distance through which the valves rotate is large.

The best way of restricting access is, of course, to shut the valves. This deters predators that invade by way of the aperture, and enables some clams to survive passage through the digestive systems of shell-swallowers. Nut clams (Nuculidae) are reported to resume normal activity after spending 2 weeks inside the sea star *Astropecten*.

Predators have available to them two ways of forcing the valves apart. One is to pull against the force of the adductor muscles. Sea stars and octopuses, for example, can place suckered arms on each valve and pull them apart. Another way is to shear the valves, pushing the two valves in opposite directions past each other in the plane of contact between them.

The resistance a clam can offer to having its valves pulled apart depends on the cross-sectional area of its adductor muscles. This area can be gauged from the size of the impressions where the adductors are inserted on the inner valve surfaces. Resistance also depends on where the adductors are located relative to the shell's hinge. Large adductors situated centrally on the valves should provide the best protection against predators that pull valves apart.

Several modifications of the valve margins reduce shear. Most obvious are the hinge teeth, located dorsally beneath the umbo, and serrations or crenulations along the free margins of the valves (fig. 4.12). These features may be so large that some protection against shearing is provided even when the valves gape slightly.

Most marine clams have well-developed hinges. Among the 2% of marine species that do not, most are deep burrowers (the lucinid *Anodontia*, for example) or live in the deep sea. Hinge teeth are lacking in 6%–17% of freshwater clams, depending on which assemblage is being examined. Surprisingly, perhaps, this absence is not compensated for by unusually large adductors, strong ligaments, or a crenulated valve margin.

Valve crenulation also occurs very widely. It is especially common among shallow-burrowing species and among those attached to rocks. Deep-burrowing clams generally have smooth valve margins that do not meet at all points when the shell is closed. These gaping clams, in which access to the shell is never fully restricted, have weakly developed hinges and small adductor-muscle impressions.

Crenulated and gaping clams show complementary distributions. Among hard-bottom clams, crenulated margins occur in 50%–57% of species in the tropical western Pacific, 35%–55% of species in tropical America, and 16%–33% in cool-temperate northern species. In clams living in sand and mud, there are no distinct differences in the incidence of crenulated margins among tropical regions, but the incidence in the tropics (35%–50%) is higher than in cool-temperate waters (23%–30%). No crenulated margins are known in freshwater clams. Permanently gaping species are common where those with crenulated edges are rare. They make up less than 16% of most tropical assemblages of burrowers, and 22%–31% of comparable cool-temperate faunas. No clams from firm bottoms are permanent gapers. In fresh water, 25%–46% of species have a gape somewhere along the shell margin when the valves are shut.

The Shell as an Offensive Weapon

We tend to think of shells as providing passive resistance to predators, but there may be instances in which molluscs use their hard parts in combat. As usual for many aspects of shell function, there has been very little work on how shells might be used as offensive weapons; but several characteristics of shells appear as if they could be used effectively in combat.

In chapter 5, I pointed out that *Acanthina* and several other drilling muricids use the downwardly projecting spine on the outer lip as a wedge or chisel to open barnacles and clams (fig. 6.16). The spine is

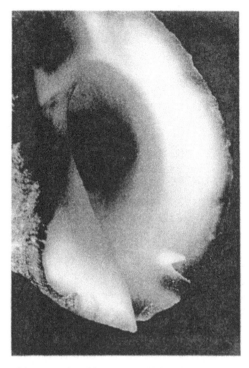

Fig. 6.16. *Acanthina monodon*, Montemar, Chile. The outer lip bears a sharp downwardly projecting tooth or spine. This specimen is 41 mm long.

often sharp, and could serve as a potent weapon against a potential intruder, but whether it is actually used in this way is unknown.

Some limpets clamp the shell down on the foot of predatory snails, and clams may occasionally close their valves on the feet of birds or the projecting parts of predatory snails. Most reports of such entrapment are anecdotal. The large Californian limpet *Lottia gigantea* uses the thickened front edge of its shell to push other limpets out of its territory on the rock. This kind of territoriality is common among large limpets and may also occur in some chitons, but it is probably rare in other molluscs.

Several tropical western Pacific sand-dwelling snails have a row of sharp spines projecting downward or forward from the edge of the outer lip. Examples include the conch *Strombus dentatus* (fig. 6.17), the cassid helmet shell genera *Casmaria* and *Phalium*, miter shells of the genus *Mitra*, and several nassariid mud snails of genera such as *Alectrion* and *Niotha*. It is possible that these spines are used in con-

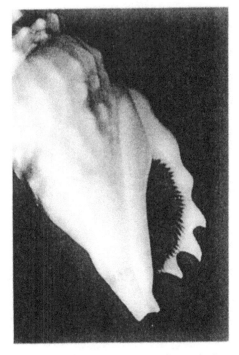

Fig. 6.17. *Strombus dentatus*, Pugua Patch Reef, Guam. A close-up of the outer lip reveals several forwardly projecting spines. This specimen is 35 mm long.

junction with rapid thrusting movements by the snail to ward off enemies. The families in which these spines occur are well known for strong and rapid movement. Strombids, cassids, and nassariids also bear opercula with sawtooth margins. I do not know whether such edges are effective against predators, but I can attest to the power with which the snails can wield the sharp opercula as the foot flails in attempts to dislodge the attacker.

Chemical Defense

A very widespread form of defense in plants and animals is the possession of substances that poison or discourage would-be attackers. Because most molluscs have external shells, chemical defenses are not so prominent in this group as in other animals. Nevertheless, lineages in which the shell became internal or was lost entirely have come to rely heavily on chemical defenses, or have gone into partnership with plants and animals that possess such substances.

Among shell-bearing snails, some muricids and volutids exude a purple dye that in principle could have an antipredatory function. Naturalists familiar with the molluscs of tropical American shores will never forget the odor of garlic associated with the purple dye exuded by muricids of the genus *Plicopurpura*. Flamingo tongues (ovulids of the genus *Cyphoma*), which live and feed on sea fans on West Indian reefs, have bright coloration on the mantle that covers the thick shell in life. Gary Rosenberg at the Philadelphia Academy of Natural Sciences thinks that these colors warn predators that the prey is distasteful. Should a predator nevertheless persist in its attack, the thick shell serves as an additional defense. We know little about how widespread warning coloration and any associated chemical defense is in molluscs.

The production of copious mucus is perhaps the most common form of defense by means other than the shell. Mucus enables limpets and chitons to cling to rocks and provides the mechanism for snails to crawl, but it can also incapacitate the jaws of predatory beetles and the fingers of human collectors. Land snails often exude copious amounts of sticky slime; so do mangrove-dwelling species of periwinkles of the genus *Littoraria*, some limpets (the West Indian *Lottia pustulata*, for example), some helmet shells (Cassidae), bubble shells (Bullidae), moon snails (Naticidae), and volutes (Volutidae). The slime champions, however, are terrestrial slugs, land snails that have carried shell loss to its absurd extreme.

Intimate Associations

Many shell-bearing molluscs have become intimately associated with plants and animals that are themselves well defended against enemies. In such protected settings, the guests often have reduced defenses of their own, relying instead on those of their host.

Among the favorite hosts of many molluscan groups are cnidarians. These animals—corals, sea anemones, sea fans, jellyfishes, and their relatives—are equipped with stinging cells (nematocysts) that effectively repel many predators searching for prey molluscs. Members of many families, including the Epitoniidae (wentletraps), Trochidae (top shells), Architectonicidae (sundials) (fig. 6.18), Ovulidae (cowrielike flamingo tongues) (fig. 6.19), and Coralliophilidae (coral snails) live and feed on cnidarian hosts. The Indo-Pacific scallop *Pedum spondyloideum* always cements itself to corals, and several boring muscles of the mytilid genera *Lithophaga* and *Fungicava* bore into the stony skeletons of living corals. Some of these guests are always found

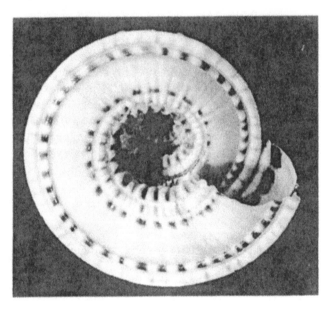

Fig. 6.18. *Architectonica perspectiva*, Nosy-Be, Madagascar. This snail lives in the sand, and is often found on poisonous zoanthids. Eggs are brooded in the broad umbilicus. Top, Lateral view of shell; bottom, ventral view showing umbilicus. This specimen is 34 mm in diameter.

Fig. 6.19. Opposite page: *Volva volva* (Ovulidae), Fly River Estuary, Papua New Guinea. This thin-shelled species lives on sea fans (gorgonians). It lacks the thickened lip of most other cowries and their relatives. Top left, Apertural view; top right, dorsal view. This specimen is 39 mm long.

Fig. 6.20. Opposite page: *Crepidula (Janacus) perforans* (Calyptraeidae), Kodiak Island, Alaska. This thin-shelled flat species lives on the inner surfaces of snail shells occupied by hermit crabs. Bottom left, Interior; bottom right, view showing that the dorsal surface is concave. This specimen is 20 mm long.

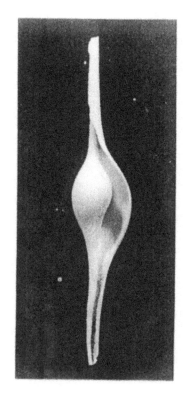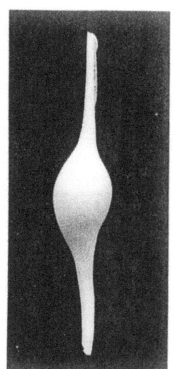

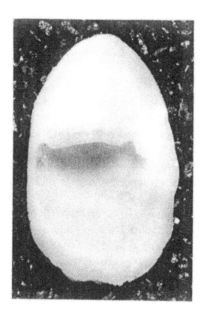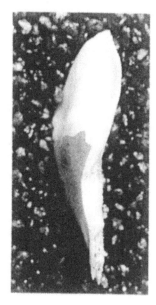

on their hosts, but others (wentletraps and sundials, for example) may spend much time in the sand as they move from host to host or simply rest. Compared to relatives that are not associated with cnidarians, these snails and clams tend to have relatively thin shells lacking some of the distinctive features of armor. Many ovulids (fig. 6.19), for example, lack teeth on the inner edges of the apertural lips. Sundials and wentletraps tend to have broadly umbilicate shells.

Two snail families, the mainly sinistral Triphoridae and the mainly dextral Cerithiopsidae, are intimately associated with sponges. Other sponge guests include North Atlantic hanleyid chitons and some keyhole limpets (Fissurellidae), which feed on their hosts, and filter-feeding oysters, vulsellid clams, and siliquariid snails. Many host sponges produce obnoxious chemicals and have barbed spicules that deter predators. Some temperate-zone scallops are typically covered with sponges, which protect them from sea stars and other predators. In the tropical western Pacific, I have often seen the trochid *Euchelus atratus* completely encased in a coating of sponge. Whether this casual association also provides protection to the gastropod is not known.

Echinoderms—sea stars, brittle stars, sea lilies, sea urchins, sand dollars, and sea cucumbers—play host to eulimid snails. Some of these small snails live inside their hosts, but others live on the outside and can move among hosts through the sand. The external guests tend to have slender, often highly polished shells, but species that are internal parasites have thin, often greatly reduced shells. Echinoderms offer particularly good protection because they tend to produce repulsive or toxic saponins, and many possess long sharp spines and other defensive structures.

The gastropod *Caledoniella* and a few small clams of the superfamily Leptonoidea live attached to the legs and bodies of mantis shrimps, whose spearlike or hammerlike feeding organs were discussed in chapter 5. How these small guests stay on their hosts when the latter's exoskeleton is molted is not known. This logistical difficulty may, however, explain why crustaceans, despite their often formidable armor and offensive weaponry, have not often become targets of host specialization by poorly armored molluscs.

Even fishes serve as hosts for some snails. In California, the nutmeg shell *Narona cooperi* (Cancellariidae) attaches itself to electric rays and sucks blood or other fluids from its host. Philippe Bouchet has described from New Caledonia a similar behavior in several sand-dwelling marginellids, which climb on sleeping parrotfishes at night and then fall off during the day. I do not know how the shells of these marine mosquitoes differ in the development of passive defenses from the shells of related species that do not associate with fish.

The only intimate associations between molluscs and freshwater hosts involve the larvae of unionid and mutelid clams. These larvae are parasitic on fishes before they fall to the bottom of a river or lake to take on the filter-feeding habits of the adults.

Many molluscs live on plants, but only a few are specialized to particular host species. Kelps, large brown seaweeds of the order Laminariales, have been common targets of specialization by temperate limpets and a few chitons. Compared to related forms that live on rocks, these species have thin, usually smooth shells. In the North Atlantic, two species of *Littorina* live predominantly or exclusively on bladderweeds (brown algae of the genera *Fucus* and *Ascophyllum*). Tropical green algae such as *Caulerpa* and *Avrainvillea* play host to several unusual snails of the order Sacoglossa. Among these, the bizarre Juliidae have evolved secondarily bivalved shells. Experiments have shown that predators are much less successful in finding snails among seaweed fronds than on rocks, and that toxic plants such as *Caulerpa* and *Avrainvillea* are usually avoided.

Finally, some molluscs live on or in other shells. Many cup-and-saucer shells and slipper limpets (Calyptraeidae) live mainly or exclusively on the shell exteriors of living molluscs and hermit crabs. If the host is nimble, the guest receives the benefit of escaping from its predators without expending its own energy. This is the case, for example, with the northeastern Pacific slipper limpet *Crepidula adunca*. When it attaches to its usual host, the trochid *Calliostoma ligatum*, the guest is safe from predatory sea stars by virtue of the rapid escape response of its host; but when *Crepidula adunca* is removed from its perch and placed on the bottom of an aquarium, sea stars find and consume it in a matter of minutes. Some slipper limpets have taken life on shells even a step further. Species of *Janacus* (fig. 6.20) typically live on the interior surfaces of shells occupied by hermit crabs. Their shells are thin and flat, and would offer little resistance if a predator could get at them. Several pyramidellid snails live with molluscan hosts and probably also receive benefits of safety from their hosts. *Ondina diaphana* on the west coast of Sweden has a delicate shell and lives inside snail shells secondarily occupied by the sipunculan worm *Phascolion strombi*.

It is perhaps not surprising that molluscs often use their shells to brood young or to attach their eggs. A large umbilicus protects the brooded young in endodontid land snails, as well as in the marine sand-dwelling sundials and the North Pacific trochid *Margarites vorti-*

cifer. Some small sand-dwelling terebrids, columbellids, and nassar-iids attach their eggs to the shell exterior. This habit also occurs in some freshwater nerites on Pacific islands. Live-bearing European species of *Littorina* use much of the shell's volume as a safe receptacle for tiny young snail progeny. The same phenomenon occurs in many small freshwater pill clams of the families Sphaeriidae and Pisidiidae.

This torrent of examples illustrates the important point that archi-tecturally inept molluscs can carry on perfectly well in settings where enemies are both numerous and powerful, as long as safe places are available. The degree of specialization to such safe places varies greatly according to locality, as does specialization in any of the other functional directions treated in this chapter. How this specialization comes to pass is the subject of the remainder of the book.

Alexander, R. M. 1990. Size, speed and buoyancy adaptations in aquatic animals. *American Zoologist* 30: 189–196. This is a crystal-clear account of the physics of buoyancy and of the various ways animals can prevent sinking to the bottom.

Coen, L. D. 1985. Shear resistance in two bivalve molluscs: role of hinges and interdigitating margins. *Journal of Zoology London* (Ser. A) 205: 479–487. This is one of the few experimental studies on how hinges of clams work.

Denny, M. W. 1988. *Biology and the Mechanics of the Wave-Swept Environment.* Princeton University Press, Princeton, N.J. Not only does this book deal with the biology of animals in the surf, but it is an excellent and clear introduction to the mechanical principles that govern form in organisms.

Garrity, S. D. 1984. Some adaptations of gastropods to physical stress on a tropical rocky shore. *Ecology* 65: 559–574. This is the best paper on the functional morphology of snails coping with the stresses of the upper shore.

Hayami, I. 1991. Living and fossil scallop shells as airfoils: an experimental study. *Paleobiology* 17: 1–18. This is probably the best, and certainly the most experimental, study of swimming scallops.

Jacobs, D. K. 1990. Sutural pattern and shell stress in *Baculites* with implications for other cephalopod shell morphologies. *Paleobiology* 16: 336–348. This is an excellent discussion of the dome principle in the design of cephalopod shells.

Lowell, R. B. 1987. Safety factors of tropical versus temperate limpet shells: multiple selection pressures on a single structure. *Evolution* 41: 638–650. An excellent account of the functional morphology of limpet shells.

Rudwick, M.J.S. 1964. The function of zigzag deflexions in the commissures of fossil brachiopods. *Palaeontology* 7: 135–171. This is an elegant functional treatment of scalloped valve edges.

Signor, P. W., III. 1982. Resolution of life habits using multiple morphologic criteria: shell form and life-mode in turritelliform gastropods. *Paleobiology* 8: 378–388. One of the few coherent functional accounts of shell form in snails.

Solem, A. 1974. *The Shell Makers: Introducing Mollusks.* Wiley, New York. This is an excellent introduction to the biology of molluscs, but it is especially strong on land snails.

Stanley, S. M. 1970. Relation of shell form to life habits of the Bivalvia (Mollusca). *Geological Society of America Memoir* 125: 1–296. This, together with the other papers by Stanley listed below, represents one of the best investigations of functional analysis in the Mollusca.

———. 1972. Functional morphology and evolution of byssally attached bivalve molluscs. *Journal of Paleontology* 46: 165–212.

———. 1975. Why clams have the shape they have: an experimental analysis of burrowing. *Paleobiology* 1: 48–58.

———. 1977. Coadaptation in the Trigoniidae, a remarkable family of burrowing bivalves. *Palaeontology* 20: 869–899.

———. 1981. Infaunal survival: alternate functions of shell ornamentation in the Bivalvia (Mollusca). *Paleobiology* 7: 384–393.

Thayer, C. W. 1975. Morphologic adaptations of benthic invertebrates to soft substrata. *Journal of Marine Research* 33: 177–289. A good introduction to the physics of how animals stay afloat on soft muddy bottoms.

Vermeij, G. J. 1973. Morphological patterns in high intertidal gastropods: adaptive strategies and their limitations. *Marine Biology* 20: 319–346.

———. 1987. *Evolution and Escalation: An Ecological History of Life.* Princeton University Press, Princeton, N.J. The book contains detailed discussions of the defenses and predators of molluscs.

Vermeij, G. J., and E. Zipser. 1986. Burrowing performance of some tropical Pacific gastropods. *Veliger* 29: 200–206.

Vermeij, G. J., E. C. Dudley, and E. Zipser. 1989. Successful and unsuccessful drilling predation in Recent pelecypods. *Veliger* 32: 266–273.

Ward, P. D. 1987. *The Natural History of Nautilus.* Allen and Unwin, Boston. An excellent analysis of the life and habits of *Nautilus*, together with interpretations of how fossil shell-bearing cephalopods worked.

West, K., A. Cohen, and M. Baron. 1991. Morphology and behavior of crabs and gastropods from Lake Tanganyika, Africa: predator-prey coevolution. *Evolution* 45: 589–607. This is the best experimental study of the predators and defenses of freshwater molluscs.

PART III

The Dimension of Time

A Historical Geography of Shells

In the fall of 1966, Egbert Leigh arrived as a new assistant professor at Princeton. As an undergraduate student, I was immediately drawn to him. Not only did his interests range far and wide, from mathematics and Chinese history to the architecture of rain-forest trees, but most importantly for me, he liked shells. As in all other matters, Leigh was unconventional in the way he arranged his collection. Traditional collectors like me arrange their shells species by species and family by family, according to an accepted scheme of molluscan classification. Leigh's system was different. All his specimens from Panama were set together on one shelf, those from the Red Sea were on another, those from Florida filled a third. This arrangement made it possible to look beyond the particulars of individual species. The assemblage as a whole became the focus of attention. Leigh's appealing approach taught me that something as simple as the arrangement of a collection subtly directs one's thinking.

Two years later, I was given the opportunity to apply Leigh's valuable lesson. Robert MacArthur was taking his graduate class in tropical ecology to Costa Rica, and he allowed fellow senior Charles Peterson and me to go along. We were treated to the full range of terrestrial environments this wonderful country has to offer, but a marine dimension was conspicuously missing. I therefore persuaded Peterson and graduate student Michael Berrill to join me on a brief excursion to Curaçao in the Netherlands Antilles, where for almost no additional expense we were able to observe and collect molluscs from that dusty Caribbean island's shores, mangroves, and desert vegetation. The emphasis was on communities of molluscs in the various habitats rather than on individual species.

Later that same year, the National Science Foundation sponsored a course on molluscs at the Hawaii Institute of Marine Biology. Thanks especially to the efforts of Alan Kohn at the University of Washington, whose seminar I had attended at Princeton, a place was arranged for me. Field excursions were an early and important component of the course. It was on one of our trips to Checker Reef, a short distance by boat from the Institute, that I realized how strikingly different the shells in Hawaii were from those at Curaçao. The cones, drupe shells, and cowries of Hawaii had apertures so long and narrow or so restricted by barriers that there hardly seemed to be any room for the soft parts inside the massive shell. Of more concern to those around me was the smell of decaying molluscan flesh—the openings were so small that effective cleaning of the shells was impossible. Even the shell exteriors were different in the two regions. Many of the Hawaiian shells were adorned with knobs and spines. All these attributes were much less well developed in the shells from comparable Caribbean habitats.

Were these differences symptomatic of a general contrast between the Caribbean and tropical Pacific, or was I being misled by the shells from two unusual and unrepresentative sites? In an effort to find out, I sought funding to collect shells from physically similar habitats in all the major tropical marine regions. In the early 1970s, funding agencies still tolerated thinly veiled fishing expeditions whose primary purpose was to uncover patterns and to make observations that might eventually serve as the basis for more directed research into underlying causes. Nowadays, it is expected that every scientist, regardless of experience, have an unambiguous testable hypothesis before any money is risked on a project. The crucial first step of making observations or identifying a phenomenon or puzzle to be explained is either ignored or ridiculed as undirected and unscientific. In 1971, however, several agencies including the National Science Foundation provided the funds I needed to carry out a geographical survey of shallow-water tropical marine molluscs. Ultimately I extended the work to cooler regions of the globe as well as to freshwater habitats. From these surveys and from laboratory studies of how shells work, a general picture of the geography and ecology of shell function began to emerge.

This work confirmed that subtle but consistent differences in shell architecture exist among physically similar habitats in distant parts of the world. Within the tropics, shells from the western Pacific and Indian oceans generally display a higher incidence and greater development of apertural barriers, strong sculpture, and many other enemy-related attributes than those in the Atlantic. Shells from the eastern Pacific occupy an intermediate position.

This pattern required an explanation. If two assemblages lived in physically similar environments in different parts of the world, why did they differ in shell architecture? Did such differences not violate the unwritten rule that similar environments elicit similar adaptive responses?

Answers to these questions must take account of history. A shell is, after all, more than just a functional structure whose form reflects the conditions of the life and death of its maker. It also bears the indelible stamp of evolutionary history, a pedigree of past events and circumstances that affected generation after generation in the long line of descent from distant ancestors. Form is, in other words, a mix of adaptation and ancestry. The same is true for assemblages of species, or faunas. Their composition must reflect present-day climate, geography, and the biological environment, as well as historical changes in these conditions. Some lineages cannot cope with change and thus become extinct; others adapt, and may even diversify through a process of evolutionary splitting, or speciation. Still others invade as immigrants and adapt to a new life after having evolved elsewhere. Finally, there are lineages that, despite all the tumultuous change around them, remain architecturally the same.

The obvious source to consult if one has questions about the history of life is the fossil record. This record enables us to trace the ancestry of living species and to piece together the architectural, ecological, and geographic histories of lineages. The greatest detail is available for the evolutionary events of the most recent geologic epochs. It is these events of the last 20 million years or so with which I shall be concerned in this chapter. After introducing the four major tropical faunas of the modern ocean, I recount the highlights of physical history of the tropics. In this geographic-historical context I then ascertain how, when, and where the differences in adaptive specialization that we see among tropical regions today came to pass. The same exercise is then attempted for the faunas of the cold seas of the north.

The Geography of Tropical Marine Life

The shallow-water marine tropics are today divided biologically into four great regions or provinces, each of which supports a rich and distinctive biota of plants and animals. The geographical limits of these regions are roughly indicated on the map in figure 7.1.

By far the largest region, supporting the richest marine biota on earth, is the Indo-West Pacific (or Indo-Pacific). It stretches from the Red Sea and the east coast of Africa eastward through Indonesia, the Philippines, New Guinea, and northern Australia to the oceanic islands of the central Pacific. Some reef-associated molluscs such as *Morula granulata*, *M. uva*, and *Cerithium nesioticum* have distributions encompassing nearly this entire region. Of course, many species have limited ranges, being confined to the Red Sea or to parts of Southeast

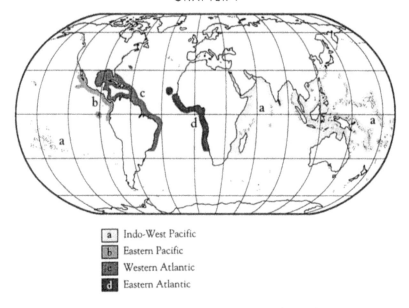

a — Indo-West Pacific
b — Eastern Pacific
c — Western Atlantic
d — Eastern Atlantic

Fig. 7.1. Major tropical marine faunal regions in the modern ocean.

Asia or to the Hawaiian Islands, but there is so much in common between the faunas of these areas that the Indo-Pacific is appropriately treated as a single faunistic unit.

Throughout the Indo-Pacific, coral reefs form a conspicuous environment that is home to an astonishing diversity of molluscs. The clear blue waters of reefs have had an almost fatal romantic appeal to biologists, with the result that more is known about reef-inhabiting species than about molluscs in less attractive environments such as mangrove swamps, sea-grass meadows, and the muddy and sandy bottoms adjacent to the continents and large continental islands. Yet these habitats also support many species. It has been estimated that there are as many as 6000 species of molluscs in the Philippines. Even such oceanic outposts as the Hawaiian Islands support about 1000 species.

Of the two biotas in tropical America, that of the western Atlantic is like the Indo-Pacific biota in miniature. There are well-developed coral reefs, mangrove communities, and sea-grass meadows, but each of these habitats supports only one-third to one-sixth the number of species seen in comparable central Indo-Pacific sites. The biota is centered on the West Indies and the Atlantic coasts of Central and South America. On the west coast of America, the tropical eastern Pacific biota extends from Baja California and the Gulf of California in the north to Peru in the south. Although some offshore-island groups such as the Galápagos are included, most of the biota is deployed along the narrow continental margin of the eastern Pacific. Reefs are small and local, and sea-grass meadows are entirely and inexplicably absent. Nevertheless, the eastern Pacific supports a very rich fauna that, except for reef-dwelling species, is generally more diverse than the western Atlantic fauna.

Despite their differences, the faunas of the western Atlantic and eastern Pacific are very closely related. Among molluscs, there are hundreds of species pairs, one member occurring in the Atlantic, the other on the Pacific side of tropical America. Where these species pairs have been investigated, the two species have been recognized as sister species, that is, as having originated from a single common ancestor. The tropical region with the smallest fauna is West Africa. In its poor development of reefs and the near absence of sea-grass meadows, West Africa is ecologically like the eastern Pacific. The tropical belt, which extends in the eastern Atlantic from Senegal to Angola, largely coincides with the West African continental shelf and with the many estuaries emptying onto it.

Historical Geography of the Tropics

The faunas of the four great tropical regions are distinct today because there are barriers between them that most species are unable to cross. The barrier between the Indo-Pacific and eastern Pacific biotas is a stretch of deep ocean that is at least 5400 km wide. A few central Pacific reef-associated molluscs are able to cross this barrier as swimming larvae, which can spend weeks to months in the plankton. Eastward transport of these larvae has enabled some species to colonize the offshore islands and sometimes even the mainland coast of Pacific tropical America. There is a similar but narrower (1900 km) barrier in the Atlantic between the faunas of Atlantic tropical America and West Africa. This barrier is apparently crossed in both directions by the planktonic larvae of snails. A land barrier separates the biotas of the tropical eastern Pacific and western Atlantic. It effectively prevents dispersal of all marine molluscs between these faunas. Even the freshwater Panama Canal, which since 1914 has provided a link for shipping between the Atlantic and Pacific, has let through only a small number of species. The barrier between the West African and Indo-Pacific regions is also a land mass. The African continent extends just far enough north and south of the tropical belt to keep species from dispersing around the Cape of Good Hope and by way of the Suez Canal, Mediterranean Sea, and northwest Africa.

We are accustomed to think of geographical features as fixed and immutable, but the geological record shows that this impression is quite erroneous. Over the span of the last 23 to 24 million years, during the time interval known as the Neogene period, some barriers to dispersal have arisen while others have disappeared. These physical changes have brought with them extinctions, alterations in the ranges of species, and opportunities for adaptive change and speciation. To understand the modern marine tropics, we must take account of these Neogene events.

Geographically, the world was a very different place 24 million years ago from what it is today. At that time, there was a more-or-less unbroken band of warm ocean stretching around the world near the equator. North and South America were separated from each other by a broad seaway; so were Africa and Europe. The continental block bearing Australia and New Guinea was as much as 15° of latitude south of its present position.

All this was to change dramatically. During the Miocene epoch, the first and longest epoch of the Neogene period, continental movements and episodes of mountain-building resulted in the formation of barriers that divided the once unbroken tropical oceans into several distinct seas. The marine connection that had existed between the Indian Ocean and the Mediterranean Sea was interrupted during the early Miocene, about 18 million years ago. Later, as Australia continued its northward trek toward Southeast Asia, the seaways between the Pacific and Indian Oceans became restricted by the rising island arcs that now comprise Indonesia and the Philippines. The land barrier is still not fully formed in this region, but continuing volcanic eruptions and massive earthquakes offer dramatic testimony that the geography of this jumble of islands between Asia and Australia is still changing.

Similar upheavals were taking place in tropical America. During the late Miocene, 10 to 5 million years ago, the seaway between the Atlantic and Pacific Oceans was becoming narrower and shallower as island arcs formed and plates carrying various land masses moved in complex ways. By the beginning of the Pliocene, about 4.9 million years ago, there were several straits still open, but by the middle of the period (3.5 to 3.0 million years ago) the marine connection was definitively broken.

As these events were taking place, the world's climate underwent substantial change. During the Middle Miocene (15–14 million years ago), and again during the early Pliocene (4–3 million years ago), warm marine waters extended well north and south of their present limits. Even in the North Sea, at a latitude of 50° to 53° north, warm-temperate molluscs prevailed during these benign spells. At other times, the limits of the warm-water faunas were pushed closer to the equator. This was especially true during the glaciations, episodes in which ice sheets covered large areas of the northern continents. These began in earnest during the late Pliocene, about 2.4 million years ago, and increased in intensity during the classical Ice Ages of the Pleistocene epoch, spanning an interval of time from 1.6 million to about 12,000 years ago.

It is not at all certain that the world as a whole was colder during glacial times. Just because the tropical belt was compressed does not necessarily mean that it was cooler. In fact, the evidence points in the opposite direction. During the early Pliocene, when warm-water organisms extended to higher latitudes than they do today, the tropical western Atlantic seems to have been 1° to 2°C cooler than it is now. The same may well have been true during the middle Miocene. Similarly, the tropics may have been a little warmer during glacial advances than today.

Water barriers between tropical marine regions are, of course, affected by climate as well as by geography and by the pattern of circulation in the oceans. Because warm water was transported to high latitudes during the middle Miocene, Early Pliocene, and other briefer warm spells, circulation at those times was much more vigorous than today. The combination of warm water and fast currents enabled some species to move between the Indian Ocean and tropical Atlantic Ocean by rounding South Africa, which today forms a physical and climatic barrier to such dispersal. Another factor contributing to rapid circulation during the Pliocene was the closure of the Central American seaway. Atlantic water that would have flowed westward into the Pacific was instead diverted to the north, greatly strengthening the Gulf Stream. This current carries warm water from the Caribbean Sea and the Gulf of Mexico northward along the east coast of the United States and then eastward across the North Atlantic to Europe. Faster currents in the Atlantic may have made it possible for the planktonic larvae of many species to cross the Atlantic, from west to east as well as from east to west. Thus, the water barrier that existed in the Atlantic was made easier to cross when circulation intensified during the early Pliocene.

In the tropical Pacific, most dispersal of larvae is from west to east in the eastward-flowing North Equatorial Countercurrent and the

Equatorial Undercurrent. Normally, these currents flow relatively sluggishly and do not reach the Pacific coast of the American mainland. The reason is that strong southeasterly tradewinds blow surface water from the west coast of South America northwestward into the tropical western Pacific. On the South American coast, this surface water is replaced from below by cold nutrient-rich water. In the west, the water carried by the southeasterlies stacks up against the land masses of Southeast Asia and the adjacent island arcs of Indonesia and the Philippines. When the southeasterlies fail, as they do every 3 to 4 years, the waters in the west return in a large slow wave eastward across the Pacific. The eastward-flowing currents speed up and reach the tropical American mainland, where warm nutrient-poor waters replace the normal food-rich waters welling up from the depths. This situation creates the so-called El Niño events, during which vast numbers of marine animals die as the result of starvation and overheating. These events typically begin in December, at the time El Niño (the Christ child) is believed to have been born. It is thought that the constriction of seaways through the Indonesian islands during the early and middle Miocene might have set up the conditions for El Niño.

Without the constriction, water pushed westward by the tradewinds would have continued to flow unimpeded into the Indian Ocean; but when the flow between the Pacific and Indian oceans was reduced as island arcs rose, water began to collect in the western Pacific. The closure of the Central American seaway probably made matters even worse, for then the easterly warm currents reaching the Americas were diverted north to California and south to Peru and Chile instead of flowing through the Panama portal to the Atlantic.

In short, the modern configuration of land masses was established by mid-Pliocene time, about 3 million years ago, but ocean currents have continued to vary in intensity. As a result, the barriers that exist among the tropical marine biotas vary in their effectiveness as climate and the pattern of oceanic circulation change with time.

Historical Perspective on Tropical Molluscs

It is one thing to recite the chronology of history—the names and dates of events, places, and participants—but quite another to understand how populations of organisms respond to change. Given the present-day architectural contrasts between the rich molluscan faunas of the western Atlantic and eastern Pacific, an ideal place to begin an inquiry into this evolutionary question is the isthmus of Panama. Today, a distance of only 50 km of hills, forests, and rich fossil deposits separates the two faunas; yet, these two faunas are today very different, and have undergone remarkably different histories since their separation 3 million years ago.

Before the uplift of the isthmus, there was a single tropical American marine fauna. Architecturally, the molluscs of this fauna resembled those of the present-day eastern Pacific. All the observed differences between the western Atlantic and eastern Pacific faunas are thus the product of the last 3 million years of history. These differences include a higher incidence in the Pacific of strong sculpture, high spires, narrow apertures, and obstructed openings in snails, a lower incidence in the Pacific of umbilicate snails, and a better representation in the Pacific of crenulated margins, burrowing-enhancing ratchet sculpture, and stability-enhancing broad truncated posterior ends in clams. Moreover, breakage seems to be a more important cause of death of snails in the eastern Pacific than in the Atlantic. Correspondingly, Atlantic crabs are smaller and have less robust claws than their eastern Pacific relatives.

Four processes could have contributed to these geographical differences. First, evolution within Atlantic lineages might have de-emphasized enemy-related traits. Second, extinction could have eliminated species with strong shell defenses and left less well defended species unaffected. Third, the splitting of lineages may have been more frequent in groups in which enemy-related traits are not well expressed. Finally, immigration of weakly defended species that evolved elsewhere may have occurred. All these processes would have reduced the expression of passive shell defenses within lineages or decreased the frequency of such defenses among species in the Atlantic fauna. As will be seen, only the first two are likely to have been important.

If there are fewer and weaker shell-breaking predators in the Atlantic today than at the time the Central American land bridge was established, traits protecting molluscs against attack by crabs might be expected to have decreased within evolving lineages, especially if the expression of such traits interfered with other functions. This may well have happened, but no well documented examples have come to light.

There is, however, a possible example to be seen in the muricid snail genus *Plicopurpura*. In the eastern Pacific, this genus is represented by *P. columellaris*, which comes in two contrasting forms (fig. 7.2). One form has an almost limpetlike shell with a very large aperture whose thin outer lip lacks denticles on its inner edge. The other form is a more tightly coiled, much thicker shell with a small aperture whose outer lip bears a series of teeth on the inner margin. In the western Atlantic, *Plicopurpura* is represented by the very similar sister species *P. patula* (fig. 7.2). This species is almost identical to the limpetlike form of *P. columellaris* from the eastern Pacific. There is, however, no equivalent to the thick-shelled narrow-apertured form. A few specimens from the Atlantic coast of Panama and Costa Rica have slightly thickened outer lips, but these shells are still thin compared to the thick form of the eastern Pacific. One interpretation of this pattern is that the common ancestor of the two species of *Plicopurpura* had the characteristics of the limpetlike form, and that the thick-shelled form arose in the eastern Pacific after the Central American barrier was formed. It is also possible, however, that the thick-lipped form was once present in the Atlantic as well as in the Pacific but became extinct in the former ocean because the predators whose attacks were effectively repelled by that form had disappeared. Unfortunately, very few fossils of the genus are known, so that we cannot be sure which of these interpretations is correct.

A second possible cause of the architectural differences between Atlantic and Pacific molluscan faunas is extinction. If species whose shells were especially well fortified were more prone to extinction than were architecturally less specialized species in the Atlantic, the incidence of such specializations would decline after the Panama isthmus formed. Studies of the extraordinarily rich fossil record of tropical American molluscs support this interpretation. After the Central American uplift, almost one-third of the genera of molluscs in the tropical Atlantic became extinct. Among the snails that disappeared, the incidence of apertural and sculptural traits that are associated with resistance to shell-breaking and invading predators was almost twice as high as it was in snails that survive to the present day. In the eastern Pacific, extinction affected only about 15% of genera living at the time the isthmus formed, and armored types were slightly less prone to it than were architecturally unspecialized forms.

Many of the groups that became extinct in the Atlantic survive today in the eastern Pacific as relicts. Some 10%–20% of species in shallow-water molluscan assemblages in the tropical eastern Pacific belong to such relict groups. Traits conferring protection from shell-breakers and invading predators are better represented in this relict group than among genera that survived from the Pliocene to the present day in the western Atlantic.

Two processes besides extinction affect the architectural makeup of faunas. One of these is speciation, the evolutionary splitting of one lineage into two or more daughter species. This process requires first that a population be subdivided into two or more isolated entities.

Fig. 7.2. Tropical American species of *Plicopurpura*. Top, *P. columellaris*, Isla Santa Cruz, Galápagos. There are two forms, a thin-shelled limpetlike form with broad aperture and simple outer lip (left), and a thick form with narrow aperture and thick, toothed outer lip (right). The average length of this species is 62 mm. Bottom left, *P. patula*, Cove Bay, Saba, West Indies. This is a thin-shelled limpetlike form with broad aperture and simple outer lip. Bottom right, *P. patula*, Fort Randolph, Atlantic coast of Panama. Though the aperture is narrower and the outer lip is somewhat thicker than in the form on the left, this Atlantic form still is a far cry from the thick eastern Pacific form. The average length of *P. patula* is 80 mm long.

This can occur if a barrier is formed where none existed before, or if several individuals from a parent population disperse to an area previously unoccupied by the species and with which contact is not maintained. Isolation by itself is not sufficient to guarantee speciation. The isolated entities must diverge from each other, that is, they must evolve their own unique characteristics so that, if the isolated groups were again to come into contact, they would behave as separate species. This means that there will be little or no exchange of genes between them.

With all the geographic and climatic changes the world has witnessed over the last few million years, one might expect isolation and divergence to have occurred frequently. The imposition of a land barrier between the Atlantic and Pacific did indeed cause populations on the two sides of tropical America to diverge, with the result that hundreds of species pairs were formed. How much additional speciation occurred within the eastern Pacific or within the western Atlantic regions has not been well studied. There are, however, many species known from one coast of tropical America that have no counterparts in the living or fossil fauna on the other side and that are not likely to have immigrated from adjacent regions. These species therefore probably evolved during the early Pleistocene, well after the formation of the land barrier between the Atlantic and Pacific. There is mounting evidence that many shallow-water sand-dwelling groups underwent substantial speciation in the eastern Pacific. Probable examples include the beach clams (Donacidae), venerid clams (Veneridae), mud snails (Nassariidae), auger shells (Terebridae), olive shells (Olividae), and dove shells (Columbellidae). For the most part, these groups have well-armored shells, and many (especially the beach clams and olives) are specialized for rapid burrowing. No such diversification of sand-dwellers took place in the western Atlantic. Instead, several rocky-shore snail groups underwent speciation in the Western Atlantic. Prominent among these are the keyhole limpets (Fissurellidae), top shells (Trochidae), and periwinkles (Littorinidae). It is striking that these limpetlike, umbilicate, and upper-shore snails are particularly characteristic of wave-swept shores, and that their shells are not notably armored or adapted for rapid locomotion. The tentative conclusion is therefore that the pattern of speciation in tropical America has contributed to the architectural differences observable between eastern Pacific and western Atlantic molluscs.

The final possibility is that the fauna of the Atlantic was invaded by immigrants that did not possess traits associated with antipredatory or locomotor specializations. Following the uplift of Central America, the western Atlantic marine fauna did receive several immigrants from West Africa and even from the tropical Indian Ocean. These species evidently crossed the Atlantic by way of the westward-flowing currents on either side of the equator. Their number, however, is small, and any change in the incidence of characteristics related to defense or locomotion would have been minor at best.

Molluscs also immigrated to the eastern Pacific, but they came from the west rather than from the east. Reef-associated snails have crossed the deep central Pacific as larvae from the eastern outposts of the Indo-West Pacific region in the Line Islands. Although these immigrants have a higher incidence of narrow or obstructed apertures than do native eastern Pacific species, they make up less than 5% of the fauna on the mainland coast of western tropical America.

The tropical Indo-West Pacific fauna has an even greater incidence and expression of traits related to armor and locomotion than does that of the eastern Pacific. It is the only region in which snails with ratchet sculpture and sand-dwelling snails with a row of spines along the edge of the outer lip are at all common. Narrow apertures occur over half the sand-dwelling snail species in the Indo-West Pacific, a higher frequency than anywhere else. Tubercles and spines also reach their peak frequencies in the Indo-Pacific among rock-dwelling snails. Breakage is a leading cause of death for snails in many parts of the western Pacific, and the predators responsible are unparalleled in their degree of specialization.

What historical circumstances set the Indo-Pacific apart from the other tropical marine regions? For one thing, extinction has done little to diminish the diversity of the fauna. Several species and genera have undergone range contractions since the Pliocene, and a few others have become wholly extinct. Most of these changes affected the oceanic islands of the central Pacific, where several species that occurred during the late Miocene and Pliocene have since become restricted to the shores of continents and large islands in the western Pacific. As far as I have been able to determine, at most five or ten genera have become extinct in the Indo-Pacific region since the Pliocene. The fauna has therefore remained largely intact for several million years while the faunas of tropical American were substantially impoverished by extinctions.

Speciation probably also contributed to the high incidence of shell specializations among Indo-Pacific molluscs. The fossil record, insofar as it has been studied, indicates that many of the familiar genera of today date back only to the Pliocene. Many of these are characterized by exceptional development of apertural protection. Examples include many genera in the Muricidae (*Morula, Drupa,* and *Drupella*), miter shells (Mitridae and Costellariidae), cowries (Cypraeidae), helmet shells (Cassidae), cones (Conidae), and conchs (Strombidae). Snails with ratchet sculpture and those with a row of sharp spines on the outer lip are mainly of Pliocene or Pleistocene origin. Evidence from the fossil record indicates that groups whose shells are effectively protected against breakage and predatory invasion have undergone more speciation than have less well protected groups. In other words, the distinctive features of shallow-water Indo-Pacific molluscs are the products of recent evolutionary events.

In contrast to the other tropical faunas, which have absorbed immigrants during the last several million years, the Indo-West Pacific fauna has not been invaded by molluscs from elsewhere. In fact, this fauna is the source for most of the immigrants to the tropical eastern Pacific and Atlantic. It is therefore evolution and speciation within Indo-Pacific lineages that is responsible for the unusual degree of architectural specialization.

The history of the tropical marine faunas therefore indicates that extinction interferes most with groups that are most specialized in passive and locomotion-related shell defenses, and that the degree of such specialization is greatest where the magnitude of extinction has been lowest. This pattern has been reinforced by speciation, which is most evident among specialized groups. Although much remains to be learned, especially about speciation, it appears from the available evidence that such isolation and divergence have been especially frequent in the Indo-Pacific and eastern Pacific regions.

The Temperate and Polar Regions

Unlike the tropical belt, which stretches in a single broad band around the world, the temperate and polar regions are developed in two latitudinally widely separated belts on either side of the equator. Northern and southern warm-temperate regions (fig. 7.3) contain molluscs that are evolutionarily closest to species found in nearby tropical regions. The faunas are therefore best thought of as impoverished versions of tropical ones, with some cool-water elements thrown in. For example, the Mediterranean region contains many species identical with or closely related to species in West Africa, as well as a large number of species with affinities to colder-water forms from northern Europe. The warm-temperate faunas of southern Africa, Australia, and New Zealand are quite distinctive, containing many evolutionarily rather isolated and probably ancient elements, but they

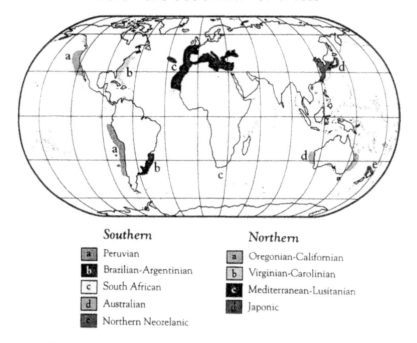

Southern

a	Peruvian
b	Brazilian-Argentinian
c	South African
d	Australian
e	Northern Neozelanic

Northern

a	Oregonian-Californian
b	Virginian-Carolinian
c	Mediterranean-Lusitanian
d	Japonic

Fig. 7.3. Major warm-temperate marine faunal regions in the modern ocean.

too have much in common with the nearby tropical faunas of the Indian and Pacific Oceans.

The world's cold marine regions bear little resemblance biologically to the tropics. If there are evolutionary links, they are generally much more remote in time than are the links between warm-temperate and tropical faunas. Figure 7.4 provides a somewhat simplified overview of the geographic division of cold marine faunas.

The shells of cold-water marine molluscs are generally much plainer and less specialized than their tropical counterparts. Moreover, whereas architectural contrasts among tropical faunas have become more apparent since early Pliocene time, today's shells from northern latitudes do not exhibit marked regional variations in architecture and have in fact become geographically more homogeneous.

During the Miocene, there were three highly distinct cool-temperate northern molluscan faunas, one in the North Pacific and two in the Arctic-Atlantic basin. In the North Pacific, a very rich fauna developed on both the American and Asian sides. So many genera and species were distributed on both sides of the Pacific that, although there were distinctive Asian and American elements among the mol-

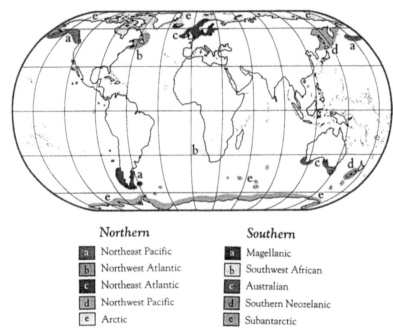

Fig. 7.4. Major cool-temperate and polar marine faunal regions in the modern ocean.

Northern		Southern	
a	Northeast Pacific	a	Magellanic
b	Northwest Atlantic	b	Southwest African
c	Northeast Atlantic	c	Australian
d	Northwest Pacific	d	Southern Neozelanic
e	Arctic	e	Subantarctic

luscs, the North Pacific formed a cool-temperate unit that had been in existence as early as 35 million years ago. The Bering land bridge separated this biota from the cold-water biotas of the eastern and western Atlantic.

Extinction and immigration during the Pliocene put an end to this Miocene status quo. A decline in the number of species began during mid-Pliocene time. The greatest losses occurred in the Atlantic, where about one-half of the North European and three-quarters of the eastern North American molluscs living during the early Pliocene have since become extinct. In the North Pacific, only one-fourth to one-third of early Pliocene molluscs have vanished. Most of these extinctions occurred during the middle and late Pliocene. The Pleistocene epoch, despite the great climatic instability produced with the repeated advance and withdrawal of ice sheets, witnessed rather little additional species loss in the sea.

Against this background of extinction, several barriers to dispersal disappeared. The Bering land bridge was transformed into a narrow seaway 3–4 million years ago during the early Pliocene. This event had a great bearing on biological conditions in the Atlantic. At least 275 species of molluscs extended their ranges from the North Pacific into the Arctic and Atlantic oceans, whereas at most 30 species invaded in the opposite direction. Pacific immigrants and their descendants comprise 10%–22% of cool-temperate faunas in different parts of the North Atlantic, but in some habitats this percentage is very much higher. On the rocky shores of New England and southeastern Canada, for example, more than 80% of the molluscan species belong to lineages that immigrated from the North Pacific during the Pliocene and Pleistocene. Included among the immigrants are such abundant species as the blue mussel (*Mytilus edulis*), horse mussel (*Modiolus modiolus*), dog whelk (*Nucella lapillus*), and the three common periwinkles *Littorina littorea*, *L. saxatilis*, and *L. obtusata*. In the North Pacific, by contrast, immigrants from the Atlantic account for less than 2% of molluscan species, and never comprise more than 10% of local assemblages.

At the same time the barrier between the Pacific and Atlantic was being broken down, currents in the North Atlantic were becoming more vigorous, possibly as a consequence of the formation of the Central American isthmus. As many as 100 molluscan species extended their ranges from the European to the American side of the North Atlantic. Still other molluscs previously confined to the warm waters of southern Europe and the Mediterranean extended northward into cool-temperate seas. A few species were even able to cross the equator, coming to Europe from southern Africa or invading in the opposite direction. Similar trans-equatorial invasions of temperate species were taking place in the eastern Pacific during the Pliocene and early Pleistocene.

These changes had the effect of homogenizing the cold-water northern faunas. During the Miocene, passive shell defenses such as crenulated valve margins in clams and narrow or obstructed apertures in snails were more frequent in the Atlantic than in the North Pacific. Extinctions affected these well-defended molluscs more than they affected other kinds of species. Moreover, immigration from the Pacific to the Atlantic further reduced the incidence of passive shell defenses, and led to an increase in the incidence of gaping shells among clams. By the late Pliocene, the temperate faunas of the North Atlantic and North Pacific had become architecturally similar.

As in the tropics, speciation in the temperate zones has been little studied. Work in Japan, California, and Europe is beginning to reveal an interesting pattern. Whereas almost no speciation seems to be taking place in the cold waters of the North Pacific and North Atlantic, many shallow-water molluscan groups underwent speciation in the early to middle Pleistocene on warm-temperate shores, particularly in southern Japan, southern California and adjacent northern Mexico, and southwestern Europe. The reasons for this proliferation remain obscure and unstudied.

What happened on the other side of the globe, in the temperate and polar zones of the southern hemisphere, is much less clear. The history of the cold faunas, which have apparently existed for at least 50 million years, is documented by only a few scattered fossil deposits in New Zealand, Antarctica, and southern South America. There is reason to believe that the modern faunas of the far south are architecturally homogeneous and are composed of many species ranging essentially around the whole of Antarctica. During and before the Miocene, there may have been more barriers to dispersal among the cold faunas.

Although the northern and southern cold regions are separated by a wide band of tropical water, there is evidence for faunal connections between them. Many molluscan groups are found in the northern and southern cold zones but not in the intervening tropics. For example, the turrid snail genus *Aforia* occurs only in the North Pacific and around Antarctica. The high-spired mathildid snail genus *Turritellopsis* occurs in both the North Pacific and North Atlantic as well as in the cold southern hemisphere, but not in between. In other instances, a southern group can be identified as the sister group to a northern one. This is the case, for example, with the periwinkle genus *Pellilitorina*, which has common ancestry with the northern-hemisphere genus *Lacuna*. These relationships imply that, although many molluscs have adapted to cold conditions independently in the two hemispheres, there has also been some communication between them. In some cases, this contact dates back tens of millions of years, but in others it is recent. The North Pacific snail genus *Fusitriton*, for example, probably reached the southern hemisphere by way of the west coast of the Americas during the Pliocene or late Miocene.

Extinction and Invasion

By now the reader should be quite convinced that the world's climate and geography have changed dramatically over the course of the last two dozen million years, and that the limits of distribution of species are anything but fixed. Many species became regionally extinct and are restricted to geographic refuges, while others extended their distributions as barriers to dispersal fell away. These changes in distribution, together with the extinctions of many species, had a profound effect on the architectural composition of molluscan faunas throughout the world. In the tropics, the erection of barriers has resulted in the exaggeration of architectural differences, whereas in the temperate zones there has been a trend toward architectural uniformity among regions, thanks to the elimination of barriers. Given the role that extinction and invasion have had in modifying the architectural composition of molluscan faunas, we should ask why extinction has affected some faunas more than it did others, and why some regions absorbed a large influx of immigrants while others received very few.

Tropical America provides some valuable clues to the question of why extinction hit so hard in the Atlantic. As mentioned earlier, many molluscs that disappeared from the Atlantic during the Pliocene and early Pleistocene still survive in the eastern Pacific. The number of these relics (110 by my latest estimate) is nearly five times greater than the number of tropical American molluscs that became restricted to the western Atlantic after having become extinct in the Pacific. At least 23 species that occurred throughout the tropical western Atlantic during the Pliocene have since become restricted to the coastal waters of continental northern and eastern South America. Were it not for these geographical sanctuaries along the continental margins of the Americas, extinction in the New World tropics would have been even more severe than it was.

These refuges have in common an abundance of nutrients and a rich supply of plankton. Compared with the classic blue waters of the island Caribbean, the coastal waters of the eastern Pacific and of Atlantic South America are turbid. Various lines of evidence point to the possibility that nutrients and plankton abundances declined after the Pliocene in many parts of the Atlantic including Florida and the West Indies, whereas they remained high on the Atlantic coast of South America and in the eastern Pacific. This decline may have been responsible for the observed extinctions. Just how the extinctions actually came to pass, and why they would have been especially frequent among architecturally specialized molluscs, is unclear. It may be that young stages were especially vulnerable to a reduction in the availability of nutrients, or that adults suffered repeated reproductive failure even if they themselves were able to remain alive.

West Africa is also a tropical refuge for molluscs. Many species known from the fossil record in the Mediterranean region are now found only in West Africa. A few genera such as *Purpurellus* (Muricidae) and *Harpa* (Harpidae) are now found in West Africa and the eastern Pacific, but are known as fossils in the intervening western Atlantic. Like the eastern Pacific, West Africa's status as a sanctuary appears to be linked to its nutrient-rich coastal waters.

These interpretations are consistent with what happened in the Indo-Pacific. Although no genera disappeared completely from this region, many Indo-Pacific molluscs underwent a substantial reduction in range, becoming extinct in many archipelagos of oceanic islands in the central and western Pacific. The coastal areas of Australia, Asia, and the Indonesian islands to which these species became restricted are rich in nutrients and plankton, and thus have the same characteristics as do the geographical refuges elsewhere in the tropics.

In the cold northern oceans, there are geographic refuges in the northwestern Pacific and northwestern Atlantic. About 20 molluscs, which during Miocene and Pliocene times lived on both sides of the North Pacific, have since become restricted to the coasts of northeast Asia on the western side of the ocean. In the Atlantic, at least 25 molluscs that are known as Pliocene or Pleistocene fossils in Europe are now found only on cool-temperate shores of the United States and Canada. Like their tropical counterparts, these refuges are or were sites of very high productivity.

Although these changes in distribution imply a decline in waterborne nutrients was responsible for the loss of species, the lowering of sea-surface temperatures brought on by the advent of ice-sheet formation at the high northern latitudes may also have played a part. Temperatures in the tropical Atlantic at the peaks of glaciation during the late Pleistocene were probably lower than in the much larger and more buffered Pacific. If tropical animals are intolerant of cool conditions, they would have been at exceptionally high risk in the Atlantic. It is striking, however, that the geographical refuges of tropical America and West Africa are cooler than are parts of the tropical Atlantic where the loss of species was greater. Upwelling of cold nutrient-rich water accounts for these slightly cooler conditions. During the early Pliocene, upwelling in the tropical Atlantic was geographically more widespread than it is today, at temperate as well as tropical latitudes; and seawater temperatures in the tropical belt were perhaps 1° to 2°C cooler than today. Cooling certainly resulted in the equatorward retreat of many species after the warm spell of the mid-Pliocene, but I am inclined to believe that many of the Pliocene and Pleistocene extinctions are linked to a decline in the abundance of plankton and the water-borne nutrients on which plankton and many molluscs depend.

Extinction in a fauna apparently makes that fauna vulnerable to invasion by immigrants. Among the world's marine regions, none has been affected by extinction since the early Pliocene more than the temperate western Atlantic, which lost about three-quarters of its early Pliocene molluscan species. Since the first pulse of extinctions, this region has received hundreds of immigrants from the North Pacific, northern Europe, and West Africa. The tropical and temperate eastern Pacific suffered much less impoverishment, and the contingent of invaders filling the voids left by the extinctions has been small. The area whose species have proved most resistant to extinction is the tropical Indo-West Pacific. Almost no immigrants have become established there since the Pliocene; instead, the Indo-Pacific has exported invading species to other parts of the marine tropics.

We can perhaps draw a loose analogy between the history of the world's marine biotas and the history of nation states. In the Miocene and early Pliocene, there were two great tropical biotas, one centered in tropical America, the other in the western Pacific and Indian oceans with outposts in the Mediterranean. The tropical American biota was divided in two, whereas the Indo-Pacific biota was somewhat reduced in extent but otherwise remained intact. The extinctions of the Pliocene left tropical America, especially the Atlantic sector, impoverished and vulnerable to invasion; the Indo-Pacific biota remained as the tropical superpower, having sustained few losses and providing immigrant stocks to other warm-water regions. In the north-temperate zone, the North Pacific filled the role of dominant biota. Of course, the status of biotas was not settled by conflicts and concerted planning as are the fortunes of nations, but the same attributes that make for success among nations apparently apply to biotas as well. A large geographical extent coupled with plentiful resources and the absence of crises seem to be favorable for the maintenance of biological specializations and for resistance to invasion. The conditions favorable to the evolution of specialization are the subject of the next chapter.

Evolutionary Economics: The Rise and Fall of Adaptive Themes

In the preceding chapter we were concerned with the most recent phase in the history of molluscs, a phase dominated by extinction and immigration. The major adaptive themes and specializations had long since become established. If we wish to understand how these specializations originated and then rose to prominence, we must look further back in time and take the long view of history, one embracing 50 million years of molluscan evolution.

From my undergraduate instruction in paleontology from Alfred G. Fischer, a truly great naturalist and broad thinker, the main outlines of this history were familiar to me early; but it was not until the mid-1970s that I came to think of it in adaptational terms. Propelling me in this direction was my long-standing curiosity about the rarity of left-handed and planispiral (or symmetrical) coiling in snails. In an effort to learn about the life habits of such snails, I decided first to conduct a survey of all groups with these unusual types of coiling. Two patterns surfaced early in the investigation. First, as already mentioned in chapter 2, left-handed and planispiral shells are more common in freshwater and land snails than in most marine groups. Planispiral marine snails tend to be small-shelled species living either in deep water or in association with seaweeds. The second, even more intriguing pattern is that these types of coiling were quite frequent in early snails, and that they declined after mid-Mesozoic time. Moreover, many ancient planispiral snails were much larger than those living in the sea today. Our work with predatory crabs and shell defenses was telling me that planispiral shells, which usually have an open umbilicus on both the right and left surfaces of the shell, would be especially vulnerable to breakage (see chap. 6). If such shells are common today in environments where breakage is infrequent, perhaps marine molluscs from the distant past also lived in situations where breakage was unimportant as a cause of death. Transitions from right-handed to left-handed coiling might also be more acceptable in such circumstances, particularly if they involved an intermediate umbilicate stage (see chap. 2).

As I delved further into the fossil record of shell defenses, it became clear that many of the antipredatory specializations that are so common in the marine tropics today did not rise to prominence until comparatively recent times. Before about 200 million years ago, the shells of warm-water marine molluscs were architecturally very similar to those in modern lakes, rivers, and cold seas. There was a corresponding increase in specialization by the predators that eat molluscs. Techniques of predation that involve the use of force became increasingly widespread over the course of time. The sea, in other words, became assault-water territory for many species.

Many other changes were taking place at the same time. Herbivores harvesting seaweeds on rocky shores evolved the ability to scrape and penetrate rock surfaces. Burrowers dug more deeply and rapidly into sandy and muddy bottoms. Animals with high energy requirements, including warm-blooded birds and mammals, ecologically replaced those with lower rates of metabolism, and became increasingly important as consumers.

In this chapter I trace the history of adaptive shell architecture against the backdrop of changing biological and physical conditions. First, I recount briefly the main evolutionary events of the past 550 million years. Then, I trace the history of the enemies of molluscs as well as the responses of molluscs to enemies. This history is then recast in economic terms. The principles are illustrated by animals that occupy molluscan shells after the original builders have died. Finally, I outline what the evolutionary-economic perspective reveals about our own species.

A Chronology of the History of Life

Given that molluscs form an integral part of the biosphere, their evolutionary development can hardly be understood without some appreciation for the turning points that mark the eventful history of life. From the sequence of sedimentary rocks and the fossils entombed in them, geologists have pieced together a chronology of events spanning 3500 million years of time. It is a chronology that to historians of human affairs has a familiar ring. Just as our written history reveals a parade of conquests, crises, and the rise and fall of nation states, so the history of the biosphere is one of competitive dominance, extinction crises, and the rise and decline of major groups of organisms, all played out in a world of changing climate and geography.

Life on earth began about 3500 million years ago (fig. 8.1). During the first phase, life was almost entirely bacterial in its level of organization. By perhaps 2000 million years ago, single-celled organisms capable of photosynthesis were beginning to transform the primordial atmosphere into one containing at least some free oxygen. Organisms with mineralized skeletons, however, did not appear on the scene until about 560 million years ago. At that time, a tubular mineralized fossil known as *Cloudina* appeared, together with the first signs of predation. Stefan Bengtson of the University of Uppsala has found that many tubes of *Cloudina* are pierced by small holes, which he interpreted as having been made by an unknown predator.

Eras	MYBP	Periods	Noteworthy Evolutionary Events	Major Extinction Events
Paleozoic	550	Cambrian	Earliest molluscs and other mineralized organisms	Early Cambrian
Paleozoic	505	Ordovician	Increase in diversity; first evidence of life on land	Late Ordovician
Paleozoic	440	Silurian	First vascular plants on land; first jawed fishes	Late Devonian
Paleozoic	410	Devonian	First forests; first specialized shell crushers	Late Devonian
Paleozoic	360	Carboniferous	Tropical coal forests; first winged insects; large land vertebrates	Permian-Triassic
Paleozoic	290	Permian	First warm-blooded vertebrates	Permian-Triassic
Mesozoic	245	Triassic	First dinosaurs and mammals	Late Triassic
Mesozoic	210	Jurassic	Increases in mineralized plankton; first birds	Cretaceous-Tertiary
Mesozoic	140	Cretaceous	Origin of flowering plants and social insects; major increases in diversity of marine life	Cretaceous-Tertiary
Cenozoic	65	Paleogene	Great increases in diversity of life	
Cenozoic	24	Neogene	Evolution of humans	

Fig. 8.1. Geologic time scale. MYBP is the date of the beginning of each period in millions of years before the present. Shaded areas indicate approximate time of extinction events.

The beginning of the Cambrian period of the Paleozoic era marks one of the most remarkable events in the history of life. At this time, about 550 million years ago, many separate evolutionary lines of organisms evolved mineralized hard parts. These included not only the earliest molluscs, but also brachiopods (bivalved animals also known as lamp shells), trilobites, echinoderms, perhaps vertebrates, and a host of animals that have proved difficult to classify in living phyla. Mineralized organisms were generally very small, but there were large animals with external skeletons composed of unmineralized organic materials. These included *Anomalocaris*, a bizarre one-meter-long predator with pincerlike grasping appendages. Also appearing in the Early Cambrian were burrowing animals capable of penetrating several centimeters into sand and mud. All Cambrian life was, as far as we know, marine.

The succeeding Ordovician period witnessed a three-fold increase in the number of families of marine animals. This increase, like the Early Cambrian explosion, seems to have been associated with a general rise in sea level. Cephalopods, which had originated during the Late Cambrian, were the top predators in the sea. In many warm parts of the world, reefs built by sponges, bryozoans, and various kinds of corals became large during the Middle and Late Ordovician. Research on spores and on ancient soils indicates that life on the dry land originated during the Late Ordovician, but it was not until the succeeding Silurian period that we see the first evidence of vascular land plants and of primitive land-dwelling arthropods. The Silurian also marked the appearance of jawed fishes.

The Devonian period in many ways was the acme of Paleozoic life. There were large reefs populated by a diverse array of animals. On land, vascular plants had become large enough to make the first forests. The first land amphibians appeared in latest Devonian time.

The history of life has always been punctuated by episodes of extinction. Several of these seem to have been global in scope. Global crises apparently occurred during the Early Cambrian, near or at the end of the Ordovician, and during the Late Devonian. During each of these episodes, reef-dwelling organisms were especially prone to extinction, as were single-celled phytoplanktonic organisms, which fix carbon from inorganic nutrients in the open ocean. The extinction episodes generally coincided with sudden falls in sea level, and were followed by rapid rises during which life diversified slowly at first, then more quickly. Only the Late Devonian crisis was apparently not linked to low sea level.

Following the Devonian, the Carboniferous period was a time of important evolutionary events on land. Winged insects appeared by the middle of the period, and large herbivorous land vertebrates became important toward the end. At tropical latitudes in what is now Europe and North America, forests of club mosses, horsetails, and ferns formed huge coal deposits. In the sea, reefs once again appeared during the Late Carboniferous and flourished during the succeeding Permian period.

The Permian brought the Paleozoic era to an eventful close. At its end came the greatest extinction crisis of the last half-billion years, during which fully half of the families of marine animals disappeared. Sea level reached a historic low as the world's continents drew together into a single large land mass.

The Triassic, the first of three periods of the Mesozoic era, was a time of recovery. In the sea, there was a rapid diversification of species in groups that were holdovers from the Paleozoic. On land, the first dinosaurs, mammals, and pterosaurs (flying vertebrates) appeared toward the end of the period. One or more extinction events brought the Triassic to a close, ushering in a time of great biological change.

The early phases of this change came during the Jurassic period. On land, this was the age of the great plant-eating dinosaurs. Many marine groups underwent a substantial diversification. A few Triassic animals had already evolved the capacity to burrow deeply into sediments, but the groups to which they belonged expanded greatly during Jurassic time and were joined by many others. In the plankton, several groups of single-celled organisms evolved mineralized skeletons, with the result that the abundances of silica and calcium carbonate were becoming increasingly regulated by organisms on a global scale. These and other innovations took place as sea levels generally rose; at the same time, the land masses that had assembled to form the Permian supercontinent had begun to break apart.

Momentous events occurred during the final period of the Mesozoic, the Cretaceous. Early in the period, several major groups appeared that were to have extraordinarily important roles in ecosystems later. These were the flowering plants and the earliest social insects (ants and termites). With their appearance, the diversity of life on land skyrocketed, and it is likely that biomass increased greatly as well. Evolution in the sea was less dramatic, but there too the diversity of plants and animals increased, so that by mid-Cretaceous time the number of families had exceeded the number at any previous time in earth history. Sea levels continued generally to rise, and world climate remained mild except perhaps at high southern latitudes. The end of the period was marked by yet another extinction crisis, which snuffed out the dinosaurs as well as nearly all single-celled plankton, shell-bearing cephalopods, and most reef-building animals.

This crisis, however, was only a temporary interruption. Life on land as well as in the sea diversified dramatically during the Cenozoic era. Depending on which group of organisms one might wish to emphasize, the Cenozoic could be called the age of mammals, the age of flowering plants, the age of insects, the age of birds, the age of crabs, or, of course, the age of molluscs.

The first half of the Cenozoic, known as the Paleogene period, began with the Paleocene epoch. This was a time of rapid biological recovery from the end-Cretaceous crisis, associated with a rise in sea level. The end of the Paleocene was marked by an unusual and only recently recognized extinction event that affected primarily deep-sea species. The succeeding Eocene epoch of the Paleogene was a time of unprecedented biological richness. During the middle Eocene, when sea level reached its Cenozoic peak, warm-water marine species extended as far north as Alaska and southern England. One or more extinction events came as sea levels dropped near or at the end of the Eocene. The Oligocene, the last epoch of the Paleogene, was distinctly cooler than the Eocene. As South America and Australia drifted north from their positions near Antarctica, the circum-Antarctic current was established, and deep-sea temperatures dropped from about 10°C to their modern values of 1° to 2°C.

The Paleogene gave way to the Neogene period. We are now on familiar ground, the modern biota having become established during the Miocene epoch of this period.

The Evolution of Enemies

It is likely that predation played a central role in the evolution of molluscan shells from the very beginning. I already mentioned that the earliest mineralized fossil (*Cloudina*) bears signs of having been attacked by an unidentified predator. Many Early Cambrian fossils show indications of unsuccessful attacks, implying that skeletons were at least sometimes strong enough to protect their makers against predation.

Little is known about the identity and techniques of the earliest predators of molluscs. Several arthropods had pincerlike appendages, and a number of worms were preserved whose digestive tracts contain whole shells. No Cambrian predators, however, seem to have been specialized as shell-crushers.

Cephalopods made their first appearance during the Late Cambrian. It is possible that some of them preyed on molluscs, but we have no evidence of specialized jaw structures. Repaired injuries in Ordovician brachiopods have been attributed to cephalopods. They are likely to be the work of predators, but who the culprits are may never be known.

Sea stars appeared during Early Ordovician time. Daniel Blake's careful work on these animals indicates that, although the earliest representatives were probably filter-feeders, predation may have evolved early during the history of the group. When the technique of extruding the stomach into the prey with or without the use of force appeared is difficult to judge from the fossil record of sea stars. Blake thinks that some Paleozoic sea stars may have been able to invade prey in this manner, but the Asteriidae, the modern family in which forced entry is a characteristic predation technique, did not evolve until the Jurassic period of the Mesozoic era.

Drill holes, mainly occurring in the valves of brachiopods, indicate the presence of drilling predators throughout the Paleozoic and Triassic. The identity of these predators may never be known, but it is possible that snails belonging to the family Platyceratidae were responsible for drill holes made during the Late Paleozoic and Triassic. Some members of this family have been found as parasites or predators on fossil crinoids (sea lilies), in which they made small holes.

The first predators with devices specialized for crushing the hard parts of other animals were eurypterids and lungfishes. Eurypterids were large arthropods known first from the Middle Ordovician, but the earliest probable crushers occurred during the Late Silurian or Early Devonian. At that time, species with massive pincers bearing worn teeth lived in inshore environments in Europe and North America. These pincers were about as robust as are the claws of living blue crabs (*Callinectes sapidus*) from the Atlantic coast of the United States. Blue crabs do crush molluscan shells, but they are not specialized to a molluscan diet, and their claws are much less robust than those of more specialized shell-crushing crabs. It is therefore likely that even the strongest eurypterids had relatively generalized diets.

Several groups of fishes joined the ranks of shell-crushers during the Devonian. The first to do so were marine lungfishes, which developed crushing plates in the upper and lower jaw by Early Devonian time. Various families of placoderm fishes became shell-crushers during the Middle and Late Devonian. Many of these early predators were snuffed out by the Devonian extinction crisis, but a few survived in the Late Paleozoic, and were joined at that time by at least ten families of sharklike fishes and by early bony fishes whose teeth indicated an ability to crush shell-bearing prey.

Arthropods other than eurypterids were not specialized to prey on molluscs during the Paleozoic, but some primitive Early Carboniferous mantis shrimps might have been capable of breaking or spearing thin-shelled prey just as many unspecialized mantis shrimps do today. Possibly an obscure group of crustaceans known as the Thylacocephala, which ranged from the Silurian to the Cretaceous, also had raptorial appendages capable of destroying the hard parts of other animals.

Several groups of potential shell-breakers appeared first in the Triassic and became specialized later in the Mesozoic. Calcified jaws evolved in cephalopods for the first time in the Middle Triassic, and were developed in several other lineages during the Jurassic and Cretaceous. Sharks, ray-finned fishes, clawed and clawless lobsters, and some marine reptiles had the equipment necessary to break shells during the later Triassic.

The remaining periods of the Mesozoic witnessed the evolution of many predatory groups. Asteriid sea stars, which use force to open molluscan prey, evolved during the Early Jurassic. So did crabs. Whether Mesozoic crabs ate molluscs, however, is uncertain. No Jurassic or Cretaceous crabs had claws endowed with large crushing surfaces, and no species showed a functional difference between the right and left claw as is seen in all specialized shell-crushers and shell-peelers today. Predatory snails evolved in several groups during the Early Cretaceous. Most important among these were drilling moon snails (Naticidae) and the neogastropods. Members of the latter order are predators that invade their prey by way of the aperture or envelop and suffocate the prey in the large foot; others, notably members of the Muricidae, evolved the drilling habit.

Variations of these techniques that speeded up the predation process appeared later. Anesthesia in conjunction with invasion, for example, appeared in the Ranellidae during the Late Cretaceous; and poisoning of snail prey by *Conus* originated no earlier than the Miocene epoch of the Cenozoic. Edge-drilling also appears to be a Cenozoic phenomenon. Forced entry by snails such as *Busycon, Fasciolaria,* and related forms is possibly no older than Pliocene. The use of a spine at the edge of the outer lip for prying open or hammering clams and barnacles was pioneered by muricids in the Early Miocene or perhaps the Late Oligocene. Predation by very rapidly burrowing snails also began during the Neogene period.

Specialized shell-breaking fishes and crabs appeared and rapidly diversified during the Paleocene and Eocene epochs of the Early

Cenozoic. The Paleocene marks the earliest occurrence of crabs whose right and left claws have become functionally differentiated into a crushing and a holding-slicing device. In the Eocene, shell-peeling calappid box crabs had appeared. All the major families of bony fishes with specialized shell-crushing dentition evolved during the Cenozoic. It is not known when shell-hammering stomatopods (mantis shrimps) first evolved.

This necessarily brief account of fossil predators in the sea indicates that all the major techniques of predation on shell-bearers were established very early. However, specialized shell-breakers did not appear until the Early Devonian or possibly the Late Silurian, and it was not until the later Mesozoic and Cenozoic eras that shell-breaking crustaceans with very highly specialized equipment evolved. Most of the predatory innovations of the Mesozoic and Cenozoic have involved the use of force and therefore a speeding up of the predation process. This means a greater use of energy.

The Invasion of Sediments and Rocks

Predation is not the only biological factor that changed over the course of molluscan history. Animals have played an increasingly active role in disturbing sediments, and have therefore brought about radical changes in the lives of molluscs on sandy and muddy bottoms. Bioturbation—the disturbance of sediments by animals—increased in depth and in intensity during the Early Cambrian, as well as from the Middle to the Late Ordovician and probably again from the Silurian to the Early Devonian. There was yet another increase in the Late Triassic and Early Jurassic. During each of these episodes, new animal groups invaded sediments, and the per-capita effect of animals increased. During the Early Paleozoic, most animals burrowed to a depth of at most a few centimeters in the sediment, but by the later Mesozoic and Cenozoic many burrowers disturbed sediments to a depth of 10 cm or more. Some crustaceans excavate tunnels 3 or 4 m deep. In the modern fauna, all groups that burrow shallowly have origins in the Paleozoic, whereas most groups that disturb sediments to a greater depth have origins in either the Mesozoic (tellinid clams, sea urchins, various crustaceans and gastropods, and lugworms) or the Cenozoic (gray whales, walruses, and many fishes). The trend toward deeper and more intense disturbance of sediments implies once again a greater per-capita energy investment.

The increase in bioturbation through time has had two important effects on molluscs. First, it causes animals resting on top of the sand or mud to sink beneath the surface (see chap. 4). Animals incapable of adjusting their position will eventually be smothered. Second, the activities of burrowers keep the sediment supplied with oxygen and therefore stimulate the growth of bacteria. Sediments in the absence of burrowers usually turns black and begins to smell of sulfur as all the free oxygen is used up. Nutrients that enter this oxygen-starved zone are out of reach of most animals, which require oxygen, and therefore accumulate. In the presence of burrowers, however, these nutrients are recycled into the ecosystem and support a wide variety of plants and animals.

A remarkably similar history is portrayed by organisms that excavate depressions or tunnels in rocks and shells. During the Paleozoic, almost all of these so-called bioeroders made excavations less than 2 cm deep, but during the Mesozoic and Cenozoic all the animal groups that evolved the bioeroding habit were capable of making much deeper excavations. Among these deep borers are various clams including date mussels (lithophagine mytilids), angel wings (Pholadidae), and members of the Hiatellidae. Chitons and limpets that excavate home scars on rocks and shells (see chap. 4) are no older than Cretaceous; the same is true of excavating sea urchins. In short, the depth and intensity of bioerosion have increased through time in parallel with similar increases in bioturbation and in parallel with specialized forms of predation on shell-bearing animals.

The evolutionary effects of these changes on molluscs were profound. During the Early Paleozoic, there were few if any specialized predators of molluscs, and bioturbation and bioerosion affected only the uppermost layers of sediments and rocks. Specialization of predators, burrowers, and borers occurred during the middle Paleozoic and again during the later Mesozoic and Cenozoic. Given these changes, we might expect to find that antipredatory adaptations and methods to overcome sedimentary instability became increasingly frequent over the course of time in shell-bearing molluscs. In particular, there should be a trend toward stronger passive defenses of the aperture as well as the development of features associated with rapid locomotion.

Molluscan Responses

Gastropods certainly conform to these expectations. High-spired shells, which indicate the ability to withdraw the soft parts far from the apertural edge, arose by Early Ordovician or even Late Cambrian time, but they comprised only a small minority of species in warm-water assemblages during the Paleozoic. Only in Jurassic and later assemblages did their incidence rise above 30%. Shells with thickened lips also arose early (during the Late Ordovician), but no Paleozoic snails had narrowly elongated apertures or openings obstructed by teeth. Less than 2% of species in Triassic warm-water assemblages had these modified apertures. The incidence of narrow apertures rose to about 10% in the Late Cretaceous, but modern tropical values were not established until the Miocene.

The only kind of apertural barrier that was at all common during the Paleozoic was a rigid calcareous operculum. Univalves with such mineralized doors appeared during the Early Ordovician and occurred throughout the remainder of the Paleozoic, Mesozoic, and Cenozoic. Attributes that make shells vulnerable to breakage have decreased in frequency among warm-water snails through time. Assemblages of Triassic and older age are characterized by a high incidence (25% or more) of species with umbilicate shells. No Jurassic or younger warm-water assemblage has an incidence of umbilicate species higher than 20%, and incidences of 10% or less are known only from the Eocene onward. Snail shells with open coiling, in which adjacent whorls do not touch, are found rarely today, mainly in the deep sea and among cemented species, but in the Paleozoic they were quite widespread.

These trends are not due merely to the evolution of one or a few groups that happen to have the attribute in question. Instead, they take shape because many groups evolved such attributes as narrow apertures, high spires, and calcareous doors independently. For example, shells whose thick outer lip is bordered within by teeth evolved among cowries (Cypraeidae) during the Late Jurassic; in dove shells (Columbellidae), helmet shells (Cassidae), and marginellids during the Eocene, and in mud snails (Nassariidae) and miters (Mitridae) during the Miocene.

The increasing emphasis on passive shell armor in snails is reflected in the frequency of repaired shell injuries. Reliable estimates of the incidence of repaired damage require large samples of extremely well preserved shells. Surface features must be so well preserved that a repaired break can be distinguished from a normal growth line as well as from a crack resulting from deformation of the shell after burial. It is hard to find large fossil assemblages that meet this standard, particularly in the Paleozoic. Nevertheless, David Schindel, Edith Zipser, and I were able to study repair in large collections from the Late Carboniferous of Texas, the Late Triassic of Italy, the Late Cretaceous of the southeastern United States, and the Late Miocene of Panama. Our analysis showed that there was a large increase in the frequency of repaired damage from the Late Triassic to the Late Cretaceous. No differences existed between the Carboniferous and Triassic assemblages, nor among those from the Cretaceous, Miocene, and the modern tropical Pacific.

The characteristics that make shells good fortresses are generally incompatible with speed. It might therefore be expected that the trend toward apertural protection, involving restricted entry to the shell, would preclude the widespread evolution of high locomotor performance. Yet, this is not the case. Many early Paleozoic snails were slow-moving or even sedentary animals. Their shells were often thick, but the only specialized apertural protection was the calcareous operculum in members of the Macluritidae and Oriostomatidae. In assemblages from fine sediments, such slow-moving snails accounted for up to 30% of the resident species. It was not until the Silurian or Early Devonian that the first motile diggers evolved. At that time, several members of the Bellerophontidae and Anomphalidae evolved shells that were covered in life by extensions of the body. If these extensions were part of the foot, these snails could have been burrowers comparable to today's moon snails, button shells, and olives. A few high-spired Murchisoniidae and Subulitidae and early opisthobranchs of the Devonian and Carboniferous, with well-streamlined external shells, may also have been burrowers. Not more than 10% of species in Paleozoic and Triassic assemblages of warm-water snails in sandy and muddy habitats were active burrowers, and none had ratchet sculpture. By the Late Cretaceous, at least 37% of warm-water snails in the very rich faunas of the southeastern United States were probably capable of burrowing in sediments. The increase of the Mesozoic continued into the Cenozoic. Exceptionally fast burrowers—many moon snails, olives, and mud snails, for example—evolved during the Early Cenozoic. At least some of these fast diggers combine a streamlined narrow-apertured shell with a large foot, which spreads out over the whole shell or over the shell's ventral surface when the snail is active. The foot can withdraw into the narrow opening when danger threatens. The usual incompatibility between passive protection and high speed is thus partially overcome in these snails.

The architectural history of clams broadly parallels that of snails. Features associated with passive protection and speed have become more specialized and more common, whereas those that combine a sedentary life-style with incomplete protection by the shell have declined and become restricted to a few safe environments. For example, tight closure, as indicated by the presence of crenulated valve margins, occurred in less than 5% of Paleozoic burrowing clams. By the Mesozoic era, the incidence had climbed to 13%–27%, but it was not until our own Cenozoic era that the incidence in warm-water assemblages exceeded 30%. Gaping shows the opposite trend. In the Early Paleozoic, the bivalved rostroconchs had permanent gapes between the valves even when the shell was shut. Warm-water Late Paleozoic and Mesozoic assemblages of burrowing clams contained a high percentage of gaping species (20%–32%). During the Cenozoic, the incidence had generally fallen below 20%.

The bivalve fauna of sands and muds today is dominated by actively burrowing species, but this was not so in the past. In the early Paleozoic, some 40% of bivalves were partially buried and were attached by a byssus to grains of sediment. In addition, there were species lying free on the sediment surface, as well as clams attached by a byssus on the sand surface and species that were anchored by virtue of greatly thickened umbones buried in the sediment. The earliest clams that can be regarded as fast burrowers were of Silurian age. They had the umbones behind the middle of the valves, and are therefore thought to have had a large foot appropriate for rapid digging. Radial and concentric sculpture that stabilizes burrowers in the sediment was rare throughout the Paleozoic, being developed in a few shallow-water genera such as *Astartella* from the Carboniferous and Permian periods. The cemented habit evolved during the Late Devonian and was acquired independently in many lineages of clams subsequently. During the Mesozoic, many of the Paleozoic architectural themes thrived, including clams that passively lay on the surface of sands and muds. In addition, however, burrowers became increasingly common, especially during the Cretaceous. Ratchet sculpture appeared in several Mesozoic groups, and sediment-stabilizing sculpture became more frequent. Fast-burrowing beach clams (Donacidae), tellins (Tellinidae), and razor clams (Solenidae and Cultellidae) are known from Cretaceous time onward. In the Cenozoic, only a few clams retained the habit of lying free on the sediment surface. Swimming clams might have existed already in the Carboniferous period of the Paleozoic, but no Paleozoic or Mesozoic clams were as highly specialized to the swimming habit as are members of the modern scallop genera *Amussium* and *Placopecten*, which did not evolve until the Miocene.

From time to time, bivalves have evolved a close partnership with single-celled plantlike dinoflagellates to produce a kind of giant superanimal that combines the ability to filter food out of the water with photosynthesis. The architecture of successive groups with this habit mirrors the adaptive changes seen in clams generally.

The earliest clams thought to have been associated with algal cells are members of the Alatoconchidae, a family that has only recently been recognized by paleontologists. In the Permian period of the Paleozoic, these huge clams lay immobile on sediments. The valves had winglike expansions of very thin, possibly translucent shell draped over the sediment, in which the mantle tissues could have soaked up light.

Another group of giants, belonging to the order Hippuritoida and generally known as rudists, may have evolved the photosynthetic habit in the Late Jurassic. These reef-forming clams, which persisted until the end of the Cretaceous, were either cemented (often to each other) or lay free. In many of them, the small left valve formed a thin lid over the massive right valve, which was cup- or horn-shaped; light may have been able to penetrate through the lid. In others, tissue was probably not covered by the shell at all, and was thus directly exposed to light.

The Cenozoic versions of the photosynthetic giants belong to the family Tridacnidae of the Indo-Pacific. This group may have evolved as early as the Eocene. Most species are attached by a massive byssus and lie with the valves widely gaping, exposing the mantle tissues to light. The valves are, however, capable of being shut. One species (*Tridacna crocea*) has become a bioeroder, and species of *Hippopus* have massive shells lying free on hard sand.

Passive protection by a rigid shell has been a dominant evolutionary theme in clams and snails from the time these groups originated in the Cambrian to the present day. Increases in locomotor performance, though substantial in groups such as swimming scallops and burrowing beach clams and moon snails, have usually been accomplished in the context of an external shell. The situation in cephalopods was different. Increases in speed were compatible with an external shell in most groups from the Cambrian to the Cretaceous, but in the long run reliance on passive protection provided by a gas-filled shell gave way to defenses compatible with a highly mobile way of life unencumbered by a shell.

Early cephalopods can best be described as weakly constructed and slow. The first cephalopods of the Late Cambrian and Early Ordovician had straight or gently curved shells whose apex pointed upward because of the buoyant gas within. The internal septa were simple and set close together. As a result, the shell was neither well streamlined nor well buttressed.

An innovation that appeared during the Early Ordovician was the development of internal deposits of calcium carbonate in the apical end of the shell. These heavy infillings enabled the shell to take on a horizontal rather than a vertical orientation, and therefore potentially decreased drag, but it did so as a cost of greater resistance to changes in velocity. In other words, the shell may have have been better streamlined, but it became cumbersome and was not well designed for high speed or maneuverability. Nevertheless, weighted straight or curved shells were the rule among Ordovician and Silurian cephalopods, and some became extremely large, achieving a length of perhaps as much as 10 m.

Strong sculpture and narrow or constricted adult apertures made their debut among cephalopods in the Early Ordovician. Both features remained rare during the Ordovician, but became frequent during the Silurian and Devonian periods. Calcareous operculumlike covers (aptychi) evolved among ammonites in the Early Jurassic and persisted until the demise of that group in the latest Cretaceous.

Despite the early success of cephalopods with a weight-and-see attitude, several shell designs providing greater potential speed and maneuverability appeared near the beginning of the history of these marine predators. Already by the Early Ordovician, a few cephalopods had coiled shells. Coiling enabled cephalopods to hold the shell in any of several orientations without the use of heavy internal deposits. Most of the early forms were loosely coiled, and many had straight body chambers. Such forms accounted for 8%–15% of the genera from the Early Ordovician to the middle Silurian, but thereafter they dwindled to less than 5% of the cephalopod fauna for the remainder of the Paleozoic. Loosely and irregularly coiled forms, many with straight body chambers, reappeared in the Mesozoic and became quite common in the Late Cretaceous. Unlike the Paleozoic forms, which were probably bottom-dwelling animals, many of the loosely and partially coiled Cretaceous cephalopods may have been passive midwater drifters.

After coiled shells achieved a majority among cephalopods during the Devonian, many lineages developed complex septa whose line of contact with the outer shell wall often were highly fluted and frilled. Coiled shells of this architecture were lightweight structures of considerable strength. They could be adapted for high maneuverability and high speed as well as for withstanding rapid changes in pressure and for attacks by predators. During the Jurassic and Cretaceous, many coiled cephalopods became laterally flattened, highly streamlined animals, whose flat shell surfaces were well buttressed from within by complex septa. By the Late Cretaceous, the only regularly coiled cephalopods were these sorts of streamlined ammonites. All strongly sculptured species with more robust, less streamlined shells had become extinct in the early part of the Late Cretaceous. The line leading to modern *Nautilus* survived, but unlike the streamlined ammonites of the latest Cretaceous, *Nautilus* is a slow swimmer with a relatively thick shell in which the septa are simple. Evidently, the conflicting demands of shell strength and fast or maneuverable swimming could not be satisfied in an animal with a chambered external shell. The theme of passive armor that had pervaded cephalopod evolution for most of the Paleozoic and Mesozoic eras was discontinued, while the groups with internal shells or lacking shells altogether persisted and thrived.

The Economics of Specialization

Now that we have a description of what happened architecturally to molluscs and some of their enemies, we can begin to inquire about the conditions that make specialization possible. From chapter 7 we know that these conditions are not always present. The early Pliocene may have been the last time that significant evolutionary specialization and diversification took place in marine molluscs. We must therefore look to the state of the world during the early Pliocene and at other times such as the Early Cambrian and the Cretaceous to identify the conditions that are favorable for large-scale adaptive evolution.

The search for these favorable circumstances is aided by applying some principles from population biology. Throughout this book we have seen that most beneficial traits are not universally advantageous to individuals. Instead, they come with costs because they interfere with other functions. An increase in shell thickness in response to an increasingly potent shell-breaker, for example, usually entails slower growth and a decline in locomotor performance. An increase in conspicuousness has reproductive benefits because it enables individuals to attract mates, but the price tag is in the form of enemies that may also be attracted. In short, organisms are functional compromises, and innovations or elaborations that confer benefits in one function can succeed only if they provide a net advantage to an individual's survival and reproductive success.

This rather severe requirement is not easily satisfied. As a result, many lineages simply cannot cope with change because they cannot respond adaptively to it. Failure to change may result in extinction or, if some members of the population have the good fortune to live in places that remain unaffected, in a contraction of range. Adaptive response may in many instances be less likely than is the perpetuation of the adaptive status quo.

If innovations did not always interfere with other important functions, they could be more easily fixed in a population. Incompatibility between functions is enforced by limitations on population growth. At most times, populations cannot grow; they are in a state of stable or declining numbers of individuals. A combination of enemies, limited resources, and the vagaries of the physical environment keep populations in check. Reproduction enables the average individual to replace itself, but long-term increases in population size do not occur. If populations are able to grow, however, the incompatibilities and compromises that usually prevail may be relaxed. Even if an innovation interferes somewhat with other functions, it may become established if the cost is not prohibitive. Expanding populations are, in other words, more forgiving of adaptive shortcomings and of the cost of innovation. The economic situation in which the population finds itself is favorable to a wide range of architectural designs. It allows some combinations of traits to survive that would be purged under the more stringent conditions of populational stability or decline.

When do populations expand? They can do so when the supply of resources increases or when the control exercised by enemies is relaxed. The availability of resources depends ultimately on how rapidly plants can manufacture organic compounds from inorganic nutrients, and on the speed and efficiency with which this food is taken up and recycled in the community of organisms. Any change such as a rise in temperature that increases the rate of photosynthesis in plants or the rate of decay should expand the resource base. Moreover, nutrients must not be exported from the community to environments where they cannot be used by organisms. For example, if valuable nutrients sink to the bottom of the ocean and accumulate in sediments, they may ultimately form deposits of oil instead of being recycled back to organisms.

Populations also expand after crises. If there is a physical catastrophe that kills great numbers of individuals or prevents reproduction, those individuals that do weather the crisis prosper in a world of plenty. Populations of many species can increase until resources, enemies, or another crisis once again limit expansion.

Sustained population growth may be necessary for the evolution of adaptive specializations, but it is not enough. Enemy-related traits evolve in the presence of enemies. For most molluscs, these enemies are mainly predators. For top-level consumers, however, the main source of danger probably comes from competitors that are after the same resources. Predators, for example, compete among themselves for prey. Individuals able to find and dispatch prey rapidly often have a significant competitive edge over those with less metabolic energy to invest in search and subjugation. Prey can contribute to the evolution of improved food-capturing methods in predators if they have the ability to harm their assailants; that is, their evolutionary effect depends on the cost of the predator's failure relative to the benefit of its success.

Evolutionary specialization therefore requires population growth in a resource-rich environment in which predators and other enemies are also present. Competition among top predators not only elicits adaptive responses in some prey species, but causes other prey to be restricted or pushed into safe environments where the predators' effectiveness is reduced. Thus, the risk of death by predation may be indirectly responsible for the occupation of such initially predator-free environments as the dry land, fresh water, the deep layers of sediment on sandy and muddy bottoms, habitats in rocks, and the bodies of plants and animals that are themselves well protected against enemies. Subsequently, greater food production in these initially marginal habitats had a beneficial effect on the biosphere as a whole by increasing the world's biomass and by recycling nutrients that would otherwise have remained locked up in sediments.

If we apply these ideas to the history of molluscan architecture, we see that the times and sites of adaptive specialization were indeed characterized by what can best be termed economic growth. The early Pliocene was probably the last interval of time during which enemy-related specialization and speciation occurred on a large scale in at least some parts of the marine world. At this time, as well as during several intervals of the preceding Miocene epoch, upwelling and other high-nutrient conditions prevailed in many parts of the world. Together with the warm-water conditions that extended to high latitudes, this environment was favorable for the growth of large populations.

Opportunities for economic growth also existed further back in time. Coincident with the proliferation of animals with skeletons during the Early Cambrian, sea levels generally rose. As continents were inundated by shallow seas, the area available for shallow-water populations of organisms increased, enabling these populations to expand. Similar opportunities existed during the great diversification of the Ordovician, the evolutionary events of the Late Silurian and Devonian, and especially the extraordinary diversification and specialization during the Jurassic and Cretaceous periods. The evolution of productive vegetation on land in the Silurian and Devonian, and the dramatic rise of flowering plants during the Cretaceous, probably increased the world's biomass and enriched ecosystems in shallow seas as much as it did those on land. Still another factor that may have contributed to the Cretaceous diversification and architectural specialization was an episode of extraordinarily intense and widespread volcanism, beginning about 120 million years ago. As a result of this geologic upheaval, vast quantities of phosphorus and other limiting nutrients were poured into the biosphere. With mechanisms such as bioturbation in place, much of this material could be recycled instead of being lost to inanimate sediments.

The economy of life, then, is stimulated by factors that are favorable to the manufacture of food from inorganic sources, and is further greatly influenced by positive feedback systems in which organisms regulate the distribution and recycling of raw materials. Because economic growth in the presence of enemies is favorable to evolutionary specialization of all sorts, including ever more potent and rapid means of competition and defense, the world has become an increasingly risky place. Individual organisms are probably no better off in terms of survival or reproductive success than they were in the distant past, but biological success today requires greater adaptive sophistication.

The Evolutionary History of the Housing Market

The sweep of economic history of the biosphere is perhaps nowhere better illustrated than among animals that have become specialized to live in the shells of molluscs after the original builders have died. The evolution of the secondary shell-dwelling habit is a complex affair with important repercussions not only for the occupants themselves, but also for molluscs. Secondary occupation is likely to evolve only if benefits are high, costs are tolerable, and the supply of resources (shells in this case) is reliable. All these factors vary from place to place and over time, and are influenced by physical as well as biological circumstances.

Shells often persist long after the death of their makers. Although they deteriorate as bioeroders and shell-breakers take their toll, shells are ideal as portable shelters for animals that do not themselves have adequate passive protection. The same shell attributes that protect molluscs have similar benefits to secondary occupants such as hermit crabs, some amphipod and tanaidacean crustaceans, and sipunculan worms. In fact, because hermit crabs can move from shell to shell with ease, they exercise considerable control over their living quarters. Many shallow-water tropical marine hermit crabs, when offered an array of snail shells of different shapes and sizes, choose heavy shells with small openings presumably because of the protection such shells afford, even though lighter and more capacious shells would allow crabs to grow faster and to carry more eggs.

Just how important shells are as secondary shelters is dramatically illustrated on almost any temperate or tropical sea shore. At one site in Guam, I found secondary occupants in three-quarters of the shells sampled. Living snails constituted a minority of shell-bearing animals.

On cold shores, where most shells are weak, the benefits of safety in a shell may be modest. Many cold-water hermit crabs are conspicuously oversized for their shells, and cannot withdraw the body into them. Naked hermit crabs occur fairly frequently. In the tropics, where many shells are well fortified, most hermit crabs are able to withdraw completely into the shell, and often use one of the claws as a plug guarding the entrance. Those that cannot withdraw are either extremely aggressive to assailants or jump out of their shells as soon as they are caught, quickly scampering or burrowing away.

If snail shells have become more fortified through time, they may have become increasingly attractive as targets for occupation by specialized dwellers. During the Paleozoic and early Mesozoic, when most snail shells lacked apertural protection, the advantage of co-opting another organism's armor may have been small compared to what it is now in warm seas. Only when passive shell armor became widespread would life in shells offer substantially reduced risk over reliance on the shell-dweller's home-grown defenses. This would be so not only because later shells were better fortresses, but also because many of the predators that were responsible for the evolution of molluscan defenses during the late Mesozoic and Cenozoic also prey on secondary shell-dwellers.

Most groups of secondary shell occupants can, in fact, be traced back to the later Mesozoic and early Cenozoic. Hermit crabs originated during the Jurassic period of the Mesozoic, and shell-dwelling groups of amphipod and tanaidacean crustaceans appear to be no older than early Cenozoic. Only sipunculans are likely to extend back to Paleozoic time. There is evidence from as early as the Ordovician that some snail shells were inhabited by secondary occupants. They have the scrape marks on the ventral surface and other indications that the shell was dragged about not by a snail, but by a subsequent occupant. Moreover, the shells are encrusted thickly by bryozoans (moss animals) that grew beyond the edge of the original shell's lip. Such enlargement by encrusters today never occurs while the snail is still alive, but instead takes place when the shell is secondarily occupied.

Life in shells has costs as well as benefits. Secondary shell-dwellers cannot as a rule build or repair houses, so that they do not incur the costs of raw materials as molluscs do; however, most secondary occupants must move into larger quarters as they grow and as the shell deteriorates. Moving, or more precisely the threat of being evicted by a competitor, exacts an energetic cost that molluscs do not have to bear. Most hermit crabs live under a chronic housing shortage. Not only is almost every available shell occupied, sometimes by as many as three hermit crabs that work out of the aperture and holes in the wall, but the shells in which they live are often substandard. Fights over shells are therefore frequent as well as costly; evicted hermit crabs are at high risk of being eaten and may be injured in the scuffle. Animals in shells that are too small are more susceptible to predation and are at a disadvantage in growth and reproduction to individuals in roomier domiciles. Shells that are too large and cumbersome are easily taken over by larger competitors. There is therefore usually a high demand for uncorroded, intact, fresh shells that offer good protection from the hermit crabs' enemies. This demand is enforced by competition among shell-dwellers and by predation.

The shell supply is controlled largely by molluscs and their predators. Hermit crabs and other secondary shell-dwellers do not themselves tend to prey on molluscs, so that they do not have a direct effect on supply; but they do apparently track predators, so that when a shell becomes empty after predation, it can be immediately occupied and not be buried or lost. In Florida, hermit crabs have been seen waiting at sites where predatory snails consume their molluscan prey. No such interception probably occurs when the predator is a shell-crushing crab. Not only is the empty shell destroyed and uninhabitable, but the waiting hermit crab might become the predator's next victim.

The rate of shell supply may be a critical factor in determining the evolutionary feasibility of specialization by animals to the shell-dwelling habit. If the supply is erratic, fresh shells enter the population only occasionally; meanwhile, the shells in circulation deteriorate and therefore become less effective as protection to their bearers. If many fresh shells become available all at once, such as after episodes of mass mortality caused by storms, most of the shells would be lost because they cannot be captured by individuals in the population. Predation by nondestructive animals provides a ready-made mechanism to make the rate of supply of shells predictable and relatively constant. The great abundance of shell-drilling and shell-invading predators in temperate and tropical seas since the later Mesozoic has made this pattern of supply possible, and may therefore account for the resounding success of hermit crabs and other secondary shell-dwellers. Where rates of supply are lower and less predictable, the secondary shell-dwelling habit is less easily achieved even if the benefits of shell-dwelling are great. In cold seas, for example, physical factors may account for a greater proportion of snail mortality than in warmer seas. The same is true in fresh water. For many snails of lakes and streams, catastrophic floods and seasonal droughts may cause most of the mortality, with the result that there is a glut of fresh shells at some times and a dearth at others. As far as I know, there is only one secondary shell-dwelling species (a hermit crab of the genus *Clibanarius*) known from a freshwater habitat. Patricia McLaughlin recently described this species from freshwater pools in Vanuatu, an archipelago in the tropical southwestern Pacific. On land, the supply of empty shells is usually too low to sustain large populations of shell-dwellers. There are terrestrial hermit crabs, but most of these obtain marine shells washed up on the shore; the shells are then brought inland.

The problem of supply could be alleviated if shell-dwellers were able to regulate the flow of shells without interfering with those who build them. This has, in fact, been done by many deep-water and high-latitude hermit crabs. These animals have entered into specialized partnerships with sea anemones, bryozoans, and sponges that encrust shells. The encrusters settle on small snail shells inhabited by hermit crabs, and then enlarge the original shell by extending the apertural rim as a tube (fig. 8.2). The shell-dwellers thus lives mainly in a house built by the encruster. The house grows along with the occupant. This arrangement at once eliminates the risks attendant with moving house and substantially reduces the demand for fresh shells. The encrusters, moreover, offer a measure of protection from the occupants' enemies by the presence of chemical defenses and various other devices. Partnerships such as these would be especially beneficial in environments where risks to hermit crabs are high but shell supplies are low, although they need not be confined to such situations. As pointed out earlier, the fossil record indicates that partnerships between encrusters and secondary shell-dwellers have existed since at least Ordovician time.

Secondary shell-dwellers may have had an important indirect influence on molluscs and thus on the quality of their housing. The accu-

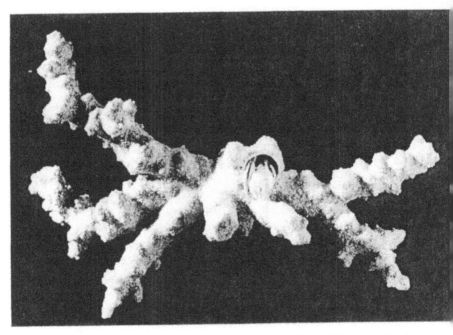

Fig. 8.2. Shell built by the West African bryozoan *Hippoporidra senegamiensis*, and occupied by a hermit crab.

mulation of cracks and the work of bioeroders cause the shells of secondary occupants to weaken with time. Even if the shell of a given species is sturdy enough to withstand attack when it is occupied by the mollusc that built it, the house may be more vulnerable under the stewardship of a hermit crab. Many shell-breaking predators go after molluscs as well as secondary shell-dwellers, and may therefore attack heavily fortified and usually unassailable shells on the off chance that the fortifications have deteriorated enough to make the attempt successful. If the frequency of such attacks on molluscs is high, there will be many opportunities during which shells are tested. Under such circumstances, additional fortification may be beneficial to molluscs. The evolution of the secondary shell-dwelling habit may have been a response to more intense predation, but itself led to a further intensification by encouraging predators to attack shells, whose number increased greatly and whose sturdiness became more variable as deteriorating shells were retained in the populations of potential prey. This is one of many examples in evolutionary economics of positive feedback, or escalation, between species and their enemies.

Molluscan History and the World Economy

The history of the biosphere is a tale of the economics of resources, of success and failure to obtain and control food, mates, and a place to live. For the last 4 million years or so, the genus *Homo* has been part of this economy, and increasingly we are dominating it as no species before us ever did. Our successes and failures to acquire goods and services may not be measured in precisely the same currency as are those that other species live by, but the principles that govern our economy are the same as those that have operated in the biosphere since life first appeared on our planet.

Economists who study our relationship with the resource supply in the present day almost by necessity have an outlook that is founded on a most unusual episode in our own history, to say nothing about the history of all of life. Even in our own limited existence on earth, the modern world economy of expansion is very young indeed, being not more than about 250 years old. Our economic expectation is predicated on the assertion that economic health means growth. Solutions to faltering economies are inevitably cast in terms of stimulating growth in jobs, resource use, and investment.

The study of the history of life contributes a unique perspective on our own economic situation. Given that the basic principles of economics—of costs and benefits, supply and demand—apply to the biosphere just as they do to our society, the fossil record provides the only truly long-term account we have of how the economic system has responded to, and been affected by, changes imposed from within and by outside circumstances. Molluscan shells offer a particularly rich chronicle of economic life and times of the past. In them are recorded the day-to-day travails and successes of their builders; their forms tell us about the risks, costs, and benefits that were important to molluscs; and by looking at shells through time, we can piece together a coherent story of changing economic conditions.

What does this molluscan perspective tell us? Perhaps most importantly, it teaches that economic growth is a rare and temporary condition. Times of expansion of populations are, for most species in most circumstances, short compared to times of stability and decline. Opportunities for adaptive specialization are similarly the exception rather than the rule. Through most of their history, species remain adaptively static, being evolutionarily constrained by enemies, limited resources, and occasional externally imposed catastrophes. It is only when populations are released from these bounds that significant alterations can reasonably be expected.

Since the dawn of the industrial revolution in the eighteenth century, the human species worldwide has experienced unprecedented economic growth. Our population has increased by a factor of five or more; our life expectancy has doubled; our share of the annually produced biomass has risen to 40%. Expansion has allowed those nations with the greatest wealth to allocate vast resources to the development and deployment of weaponry. Such resource diversion comes with costs in reduced output in other sectors, as it does in molluscs, but these costs are tolerable as long as the economy expands. Sooner or later, however, we will reach a point of resource limitation. Recycling and greater efficiency will postpone the limitation for a time, but no species or society can expect to expand and grow economically forever. Only when we have arrived at the environment's carrying capacity can we reasonably expect increases in military outlays to cease worldwide. Managing an economy that is globally in balance or even in decline will require a fundamental shift in outlook by economists and political leaders. In nature, the economy is enforced by enemies and resources; perhaps we, as a species uniquely able to plan for and predict the future, can blunt the cruelty of unfairness of competition and predation by regulating our populations and economic demand at levels well below carrying capacity.

A second lesson from molluscs and the history of life in general is that high energy use has generally been a mark of success for species living in warm, nutrient-rich environments. Energy-rich but slow-growing skeletons that are often very strong or otherwise resistant to attack have been supplanted by skeletons composed of less resistant materials that are cheaper to produce but faster to grow and repair (see chap. 3). High speed and the use of force have proved to be effective for the top consumers, among which competition to acquire resources is perhaps most intense.

A similar tendency may be discerned in our own economic development. We have come to rely on ever greater quantities of energy, both as a species and on a per-capita basis. Food production has risen because it is increasingly in the hands of large enterprises that take advantage of the economy of large scale. High energy use makes possible all manner of specializations and walks of life that would be unimaginable and unattainable with low per-capita availability of energy. This situation again closely parallels that seen in nature (see chaps. 4, 5). Only in environments where rates of biological activity are limited by low temperatures or by a low rate of supply of raw materials is high energy use not a hallmark of competitive or defensive success. Under these limiting conditions, materials that are expensive to produce and maintain—organic-rich types of shell microstructure, for example— still persist (see chap. 3). The risk of doing things slowly and efficiently are small, because enemies are less commonly encountered and are themselves metabolically limited.

There is, of course, a long-term danger inherent in high resource use. If the supply of resources is interrupted by a physical crisis, individuals dependent on a high per-capita supply face immediate starvation, reproductive failure, or both. Those with more modest energy requirements can tolerate interruptions in supply. It is difficult not to see the lessons for our own society. Industrialized nations have already had some experience with the repercussions of an interruption or a reduction in oil supplies, and wars have been fought over efforts to interfere with the acquisition of energy. The more society depends on high rates of energy use, the more vulnerable it will be to disruption if the supply is cut by enemies or by some external catastrophe.

Cooperation has often been seen as the ultimate solution to problems of resource acquisition and regulation. As we have seen in this chapter, partnerships between molluscs and photosynthesizing organisms, or between hermit crabs and shell encrusters, do make for better competitors and do result in improved control over resources; but in the end they merely raise the stakes, in the same way that the evolution of faster and stronger predators and of better protected prey raises the stakes for the hunters and the hunted. The advantages to the individuals involved are clear; but the partnerships, rather than alleviating strife, merely raise it to a higher pitch. This is not an argument against cooperation; it is simply a warning that improvements in resource acquisition come at the expense of others in a society or an ecosystem subject to resource limitation.

Shells are beautiful, fascinating, and informative works of architecture and history. Let us hope that they can also teach us how to manage the world in which their builders and we live.

Allmon, W. D., J. C. Nieh, and R. D. Norris. 1990. Drilling and peeling of turritelline gastropods since the Late Cretaceous. *Palaeontology* 33: 595–611. This is the largest published compilation of data on drilling in a molluscan group over a substantial interval of time.

Bertness, M. D. 1981a. Pattern and plasticity in tropical hermit crab growth and reproduction. *American Naturalist* 117: 754–773. This and the next two papers by one of my former students are a detailed examination of hermit-crab shell preferences, together with careful studies of how growth, reproduction, and protection vary according to the size and shape of occupied shells.

———. 1981b. Conflicting advantages in resource utilization: the hermit crab housing dilemma. *American Naturalist* 118: 432–437.

———. 1982. Shell utilization, predation pressure, and thermal stress in Panamanian hermit crabs: an interoceanic comparison. *Journal of Experimental Marine Biology and Ecology* 64: 159–187.

Briggs, J. C. 1974. *Marine Zoogeography.* McGraw Hill, New York. This provides an excellent summary of the geography of marine life.

Crowley, T. J., and J. R. North. 1991. *Paleoclimatology.* Oxford University Press, New York. Not only does this book provide an up-to-date account of what is known about climates of the past, but it explains clearly and concisely the modern climate system.

Gray, J. 1987. Evolution of the freshwater ecosystem: the fossil record. *Palaeogeography, Palaeoclimatology, Palaeoecology* 62: 1–214. Gray exhaustively reviews what is known about fossil freshwater organisms and communities, and also provides an excellent introduction to freshwater plants and animals today.

Kelley, P. H. 1989. Evolutionary trends within bivalve prey of Chesapeake Group naticid gastropods. *Historical Biology* 2: 139–156. This is one of the best studies of fossil predation.

Thayer, C. W. 1983. Sediment-mediated biological disturbance and the evolution of marine benthos. In M.J.S. Tevesz and P. L. McCall (eds.), *Biotic Interactions in Recent and Fossil Benthic Communities,* pp. 479–625. Plenum, New York. This is a comprehensive study of bioturbation and its effects through time.

Valentine, J. W., S. M. Awramik, P. W. Signor, and P. M. Sadler. 1991. The biological explosion at the Precambrian-Cambrian boundary. *Evolutionary Biology* 25: 279–356. This is the most comprehensive account of the Cambrian explosion.

Vermeij, G. J. 1987. *Evolution and Escalation: An Ecological History of Life.* Princeton University Press, Princeton, N.J. This book presents a detailed account and interpretation of the evolution of organisms and their enemies.

———. 1989. Interoceanic differences in adaptation: effects of history and productivity. *Marine Ecology Progress Series* 57: 293–305.

———. 1991. When biotas meet: understanding biotic interchange. *Science* (Washington, D.C.) 253: 1099–1104.

INDEX

References to illustrations are in italics.